Active Communication

Matthew Westra

Longview Community College

Brooks/Cole Publishing Company

I(T)P™ An International Thomson Publishing Company

Pacific Grove • Albany • Bonn • Boston • Cincinnati • Detroit • London • Madrid • Melbourne
Mexico City • New York • Paris • San Francisco • Singapore • Tokyo • Toronto • Washington

Sponsoring Editor: *Lisa Gebo*
Marketing Representative: *Marc Edwards*
Editorial Associate: *Patricia Vienneau*
Production Editor: *Tessa A. McGlasson*
Manuscript Editor: *Barbara Kimmel*
Permissions Editor: *May Clark*
Design Editor: *E. Kelly Shoemaker*

Cover and Interior Design and
 Illustration: *Design Associates, Inc.*
Advertising: *Romy Fineroff*
Typesetting: *Bookends Typesetting*
Printing and Binding: *Malloy
 Lithographing, Inc.*

For more information, contact:

BROOKS/COLE PUBLISHING COMPANY
511 Forest Lodge Road
Pacific Grove, CA 93950
USA

International Thomson Editores
Campos Eliseos 385, Piso 7
Col. Polanco
11560 México D. F. México

International Thomson Publishing—Europe
Berkshire House 168–173
High Holborn
London WC1V 7AA
England

International Thomson Publishing GmbH
Königswinterer Strasse 418
53227 Bonn
Germany

Thomas Nelson Australia
102 Dodds Street
South Melbourne, 3205
Victoria, Australia

International Thomson Publishing—Asia
221 Henderson Road
#05-10 Henderson Building
Singapore 0315

Nelson Canada
1120 Birchmount Road
Scarborough, Ontario
Canada M1K 5G4

International Thomson Publishing—Japan
Hirakawacho Kyowa Building, 3F
2-2-1 Hirakawacho
Chiyoda-ku, Tokyo 102
Japan

Printed in the United States of America

10   9   8   7   6   5

Library of Congress Cataloging-in-Publication Data

Westra, Matthew, [date]
    Active communication / Matthew Westra.
        p.   cm.
    Includes bibliographical references.
    ISBN 0-534-34007-5
        1. Public speaking.   2. Oral communication.   3. Listening.
    I. Title.
 PN4121.W353   1995
    302.2—dc20   95-21248
                                                               CIP

To my parents, Charles & Catherine,
who are my origin.

To my wife, Cheryl,
who is my present.

To my children, Benjamin & Cameron,
who are my eternity.

Contents

CHAPTER THREE

Attitudes for Active Communication - *28*

CHAPTER FOUR

Skills of the Body - *56*

CHAPTER FIVE

Probes and Questions - 71

CHAPTER SIX

Skills of Reflection - 97

CHAPTER SEVEN

Speaking to Be Heard - *112*

CHAPTER EIGHT

Applications - *137*

CONTENTS

Preface

Active communication requires more than sitting quietly while another person speaks. It is a dynamic, involved process. This book is intended to teach the skills and attitudes necessary for people to become active participants in communication, focusing on the interdisciplinary application of listening and speaking skills. Rather than presenting a scholarly review of research, I employ a conversational tone to address the needs of people who will have to act in their own and others' best interests to ensure that the messages received are the ones that are intended.

Active Communication was born out of a frustrating attempt to find a text that provides the qualities just described. While co-teaching Crisis Interviewing, a colleague and I found the text for listening skills to be inadequate, and the students shared our dissatisfaction. Many texts failed to include exercises for practicing skills, communication theory, discussion of the attitudes that benefit open communication, and material on how to create messages that will be heard accurately. Other instructors commented that they were similarly frustrated in their attempts to find a satisfactory textbook. So, we abandoned our search, and I created a manual of my own. Faculty members used the manuscript in upper and lower division courses, and both professionals and students provided feedback for revisions. This broad-based effort produced what you now hold in front of you.

Active Communication is intended for anyone who has a need or desire to improve the ability to listen and speak in a manner that facilitates accurate personal communication. Although its most frequent audience will be students in Human Services, Social Work, Counseling, and Psychology, this book is also directed toward people in Education, Speech Communication, Medicine and Allied Health, and Business. Everyone whose professional or personal lives depend on knowing and understanding the thoughts, dreams, and needs of others must be able to listen.

This book was written with the assumption that the reader has taken no prerequisite courses and has no particular psychological background and no psychological or theoretical expertise, although these may facilitate skill learning. *Active Communication* was designed to serve as a primary text in lower division Interviewing Skills courses and in upper division or graduate courses that cover communication skills within Counseling, Social Work, or Management.

In this book, I use the pyramid concept to demonstrate that each set of skills and ideas builds on previous levels, reaching toward a peak. The book begins with an overview, like a map that is studied before taking a road trip. The reader then moves from communication theory to attitudes that facilitate active communication, to listening and questioning skills, on up to helping others become better communicators—the peak of the pyramid. This approach draws readers through the several levels of active communication toward an overall experience, integrating the theory, attitudes, and skills.

Throughout the text are real-life examples and exercises that prompt readers to observe, reflect on, and practice what they are learning. Because *doing* is the best way to learn skills, the exercises offer readers opportunities to act on what they learn.

ACKNOWLEDGMENTS

No one stands on the ground, but rather on the shoulders of all who have laid the foundation for those to come. I certainly cannot claim ownership of communication theory or skills—only of the way I perceive and present them. Nor can I take sole credit for bringing the book into published form.

I would like to express my appreciation to those who have helped me with ideas, comments, and suggestions and to those who have tolerated my withdrawal from normal life. These people include, but can't possibly be limited to Mary Moline, of California State University, Fullerton; Roger Worthington, of Boston College; Paul F. Jacobson, in private practice; Gerald Corey, of California State University, Fullerton; Marianne Schneider Corey, in private practice; Cheryl J. (Samson) Westra; Patrick Callanan, in private practice; Richard McDonald; Ed Lane; Bill Parrish, of Jordan Intermediate School; Mary McMullen-Light, of Longview Community College; Jane Zeitner, of Eastern Jackson County Community College; Nathan J. Westra, for computer graphics during early versions; Paulo Flores; Theresa Bettencourt, Fitchburg State College; Mary Jo Blazek, University of

Maine at Augusta; Miriam Clubok, Ohio University; Jon Fitch, Lyndon State College; John Hancock, Fitchburg State College; Bob Innes, Vanderbilt University; and classes full of students who provided feedback during the manuscript's development. Lisa Gebo, my editor at Brooks/Cole, provided invaluable assistance by answering all my big and little questions on form, style, scheduling, and the rest. Others at Brooks/Cole who contributed to the book's production with their professional and personal talents are: Tessa McGlasson, production editor; Kelly Shoemaker, design coordinator; May Clark, permissions; and Adrienne Carter and Romy Fineroff, advertising. By her graceful copyediting, Barbara Kimmel made the book smoother and more readable.

This list cannot possibly cover everyone who has had an influence on the writing and production of *Active Communication*. To those not specifically mentioned, I hope you recognize your own part in these pages, and I thank you.

Matthew Westra

The Pyramid of Active Communication

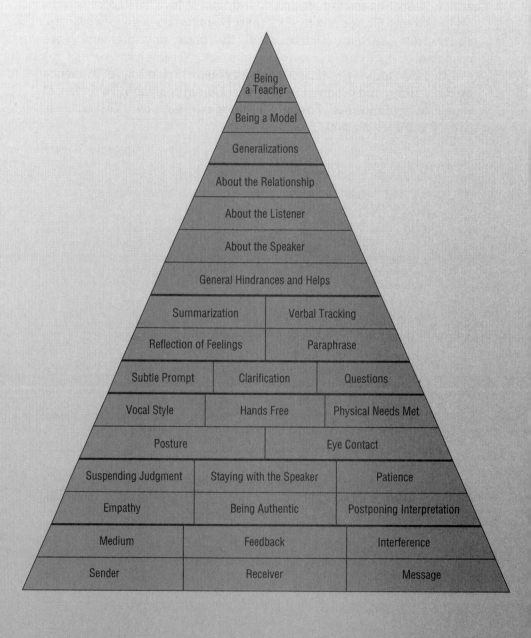

"Talking with him was such a high. He *listens* with an intensity most people have only when *talking*."

JANE WAGNER

The Search for Signs of Intelligent Life in the Universe

This chapter briefly introduces each of the coming chapters and describes how they fit into the whole pyramid of active communication. The chapter topics are Process of Communication, Attitudes for Active Communication, Skills of the Body, Probes and Questions, Skills of Reflection, Speaking to be Heard, and Applications. Exercises suggested throughout the text will provide opportunities for you to explore the concepts and practice the skills presented in the book. This chapter also includes a section on ethics. Ethics provides us a code of conduct to follow to use our skills to help people rather than to hurt or manipulate them. Ethical behavior protects both the speaker and the listener and allows productive interactions to take place. Take the ethical considerations to heart and live by them.

Active Listening: The Foundation of Good Communication

Being a good listener is the first step to becoming an active communicator. Active listening is more than simply being willing to be quiet while a friend talks. Although this willingness demonstrates an attitude of caring that is part of good listening, there is much more to helping another through listening.

Active listening is helping others organize, form, and present their thoughts. This skill is more difficult to learn than it appears. A high level of active listening requires paying concentrated attention not only to the speaker's words but also to nonverbal behaviors, feelings, patterns of consistent or inconsistent expressions, and more. At the same time, active listeners must pay attention to their own thoughts and feelings.

Paying attention to your own thoughts and feelings may sound easy, but it usually requires some conscious development. In this modern society, we are often taught to suppress and ignore our own emotions. After years of practice, most of us become rather accomplished at hiding our feelings. We lose the ability to identify what we are feeling, much less recognize and respond to these feelings while also paying attention to other people's feelings. In Chapter 3 there are exercises to redevelop this ability to tap into our emotions.

Levels of the Pyramid

This book uses the analogy of a pyramid in structuring the development of active communication skills because each level builds on those that have come before. The pyramid is divided into seven levels, each represented by a chapter in this book. Each level has been further subdivided into blocks that represent specific skills, attitudes, or portions of the concept being described. You need to attain a degree of competence at each level before advancing to the next level. This is not to say that one level must be mastered before advancing; because the levels build on one another, however, attaining competence at each level before advancing will help you understand and reach competence at the next level. It would be unrealistic to expect to have complete mastery at one level before moving on to the next. Probably no one has ever achieved perfection at any level. Just get as much understanding and practice as you can to help you advance to the next level. You will have plenty of opportunity to further develop your abilities later.

PROCESS OF COMMUNICATION

It is my belief that we cannot develop into good listeners without first understanding the process of communication itself. Despite all the pitfalls and quirks inherent in human communication, people have the remarkable ability to blend word selection, voice inflection, body language, writing, and more.

Knowing how this process works and where it goes wrong will save the listener countless hours of frustration and misunderstanding. Because understanding the process of communication is the foundation of the ability to communicate, we will examine this process in some detail.

ATTITUDES FOR ACTIVE COMMUNICATION

Successful communication requires a collection of attitudes that will put the listener in the proper frame of mind to be receptive and helpful to the speaker. I use the term *attitude* to mean a mind-set—a value system, or way of responding to and thinking about one's physical and social environments. It is more than a liking or disliking of something.

The attitudes for active communication are having empathy, being authentic, postponing interpretation, suspending judgment, staying with the speaker, and having patience. Each of these attitudes helps the listener remain open to what the speaker is saying and prevents the listener from evaluating what is being said or focusing too strongly on his or her own inner dialogue.

SKILLS OF THE BODY

Skills of the body include posture, hand gestures and placement, being comfortable, and actions such as eye contact. These physical skills are more subject to habit learning than are those of the next category.

This level of the pyramid builds on the level that precedes it; skills of the body are easier to develop when the mind-set is proper. And because these actions take care of some of the more rudimentary things that need to be done in a listening situation, skills of the body make it easier to develop the skills of reflection later on.

PROBES AND QUESTIONS

This is the first level in which the listener applies speaking skills. In any listening situation, people need to be encouraged to continue speaking and to provide information. Using probes and questions is a fine art. People often get the wrong information because the way they construct their questions actually requests the wrong information. At this level, you will learn (1) to help speakers continue, (2) to elicit specific information, and (3) to elicit a more complete narrative of the speaker's thoughts, feelings, and experience.

SKILLS OF REFLECTION

Reflecting the content and emotions that speakers express is one of the most crucial skills to help guide speakers in organizing their thoughts and feelings. Speakers occasionally comment that a therapist or other competent listener seems to have ESP. This sense that the listener knows what the speaker is thinking is generally a response to the unfamiliar experience of sensitive reflection.

The skills of reflection include reflecting feelings, paraphrasing, summarizing, and verbal tracking. With artful application of these skills, the listener can help guide the speaker to the heart of the matter without wasting time in superfluous rambling.

Although you will probably focus most of your effort between learning probes and questions and the skills of reflection, these skills are not at the peak of the pyramid. Aside from being able to listen well to others, full communication means being able to speak so that others can clearly understand. This brings us to the next level of the pyramid.

SPEAKING TO BE HEARD

Active communication generally focuses on the listening skills, which are presented first. To be a truly skilled communicator, you must listen well to what is being said and speak in a manner that helps the other person understand exactly what you mean.

Chapter 7, "Speaking to be Heard," will teach you how to speak in a way that helps prevent people from responding defensively. We will consider how the speaker, the listener, and the relationship between the two all contribute to defining the best way to construct messages. The particular skill areas discussed in this chapter are "I" statements, directness, and caring confrontation.

APPLICATIONS

The top level of the pyramid is the applications level, which is the most important one. Using these skills is proof of how well you have learned the lower levels and will help improve your relationships with people through effectively exchanging thoughts, feelings, and information. You will find plenty of opportunities to apply these skills in business, counseling, and family settings and anywhere else you happen to be. But restricting the listening skills to only professional helping or business applications is like appreciating art only in a museum.

Application also means sharing the "secrets" with others. Through teaching what you have learned about taking an active role in communicating, you will ultimately be helping others live more effectively. If others comment on your behavior when listening or speaking, tell them what you are doing. Tell them why you phrase things the way you do, and describe the skills and attitudes. Personally, I consider teaching active communication to be the peak of the pyramid. By teaching, you affect not only those you teach directly but also anyone your students affect. That is a great accomplishment.

Just a word about practice: one of the chief aspects of applying these skills is to practice them. It is essential that you practice all the skills and attitudes learned in active communication if they are to be a useful part of your ability to work with people. You may be surprised how many times in the day you will be able to practice the skills, and people will appreciate your efforts.

However, be careful not to adopt a role of "listener" or "helper" in place of your own personality, especially when you are with friends and co-workers. They will notice the role, and they may resent your acting as a "model therapist." This role implies that they have problems you are above having. Your intention may be to help them, but they may not be able to see beyond the role, and you may be out of line by assuming the role.

The Pyramid as a Whole

The pyramid of active communication is more than just a collection of skills or building blocks stacked together. Each level enhances the others, creating a whole—a gestalt in which the whole is greater than the sum of its parts. It is impossible to develop real proficiency at active communication without taking the entire pyramid into account. Skipping levels that do not seem as interesting as the others may make the process move more quickly, but the skills will be as faulty as their foundation.

I cannot overemphasize the importance of spending time familiarizing yourself with the information and skills presented at each level of the pyramid. Each level takes patience, practice, and persistence. Give yourself the time necessary to do a good job of learning, and the skills will do a good job of rewarding your effort. Don't be like the person who prayed for the gift of patience saying, "I want to be patient, and I want it NOW!"

Ethical Considerations

> **Whoever fights monsters should see to it that**
> **in the process he does not become a monster.**
> **And when you look long into an abyss,**
> **the abyss also looks into you.**
>
> FRIEDRICH NIETZSCHE

In the pages that follow, ethical concerns will be mentioned often. I believe that ethics should always be kept in mind whenever someone is trying to help another. Tough questions will arise, such as:

What do I get from helping this person?
Do I care about this person?
Do I care about this person's problem?
Do I get a payoff if the person decides to do one thing instead of the other?

Of course, there is always a payoff of some sort. Perhaps the payoff is knowing that you have done your job well. Perhaps you just feel good about yourself when you help someone else or when a client decides to use further services. Or maybe you benefit from the feeling that your job is more secure. Very small matters can get in the way of helping people for their betterment alone.

Corey, Corey, and Callanan (1993) give a good review of ethical issues for people who are in a situation to help another. Some additional ethical issues follow.

Know Yourself.   Probably the most important ethical consideration is to separate your own issues and concerns from those of the other person. To know yourself means having a sense of yourself and being able to incorporate that knowledge of self into your work and your relationships.

It helps to occasionally ask, "Who am I?" in a deeper sense. Writing down whatever you think at the moment, in a free-flowing style, can help you identify some thoughts and feelings you might not have fully articulated or explored before.

◭ *Exercises* — *Write about It*

Given the advice to explore and articulate, write out answers to the following questions, spending some time on each one. Also, be wary of responding with knee-jerk answers that might be what you have been told are the "best" type of answers.

1. How have your past experiences and relationships made you who you are today?
2. How have these experiences and relationships brought you to where you are in life today
   a. physically?
   b. socially?
   c. economically?
   d. politically?
   e. personally?
3. How have these experiences and relationships strengthened you?
4. How have these experiences and relationships weakened you?
5. How will these experiences and relationships influence you as a listener?   ◭

It is also important to examine your personal assumptions about people. This can be particularly challenging, because many of us have been taught that the only acceptable belief is that everyone is equal, deserving, worthwhile, and so on. Although this is an admirable goal, on closer inspection it becomes apparent that we all categorize people.

We prefer to spend time with some types, and we avoid other types; we respect some, and we detest some. The way we categorize is not always fair, just, right, or even rational. Don't be afraid of encountering and identifying your own preferences and prejudices.

I have yet to meet a person who does not harbor some type of generalized dislike. Personally, I have a hard time with adolescents as a group; they "push my buttons." Their arrogant belligerence makes me feel hostile toward them. Note how this statement puts the responsibility for my reaction on them, rather than on me. Also, note the use of the term *arrogant belligerence.* This label shows how I tend to generalize, and therefore expect, certain behaviors and attitudes. Of course, if we look, we'll always find something to interpret according to the label. I actually enjoy many individual adolescents, so I know this is an overgeneralized prejudice.

The benefit of identifying my response is that now I know that unless I work to resolve some of these issues, I should not work therapeutically with adolescents. Nor should I teach high school. Most likely, because of my own reactions, I will not be forthcoming, fair, or give the respect rightly deserved or earned by individual adolescents. It's important that I too know this. Now it's your turn.

⬧ *Exercises — Write about It*

Spend some time responding to each question, and when you feel done, press on a little further. Reach for your hidden reactions.

1. Are people basically good or basically bad?
2. Do people need solutions or guidance? What do these terms mean?
3. Is everyone equal, or are there differences that make some people "better" or "more deserving" than other people? What are these differences?
4. Are people's problems the result of something deep within them?
5. Are there honorable and dishonorable ways of making a living? Is there anything you would be embarrassed about or ashamed to do for a living?
6. What sorts of things should people do for recreational, spiritual, educational, and other types of fulfillment? What should they avoid?
7. What makes someone seem odd?

8. Describe someone you would want to become friends with.
9. Describe someone you would *not* want to become friends with.
10. What irritates you about people?

When you have finished, examine what you've written to see how you evaluate things, activities, and people.

What common themes do you see?
What do the things you wrote say about you?
Do you see any obvious biases that would interfere with the
    way you might work with someone?
Is there anything about your judgments that you would like to
    change?

What you wrote reflects an important part of your identity and personality. You draw from your own personal experiences, what you have been taught, and influences from the culture in which you live. Take some time to make thoughtful choices about any changes you might want to make and how to make those changes. Also, think about what you would like to keep as it is. ◬

Hidden Agendas.   The hidden agenda is any direction the listener tries to impose on the speaker but that the listener keeps secret. Hidden agendas can come from speakers as well. Whether they are meant to hurt or not, these agendas can be manipulative, passive/aggressive, and game playing. Often, the listener won't even be consciously aware of this direction or the intent to lead a speaker.

One way the hidden agenda occurs is through "leading by following." That may sound strange, but there are ways of leading a speaker by selecting to listen only to parts of what is being said. In the two examples that follow, watch how Phoebe gets her way without ever asking for it. In the first example, she doesn't want to listen to Joan; in the second, she changes the topic of conversation to one that interests her.

Phoebe would never tell Joan not to talk about the difficulties of caring for a parent with Alzheimer's disease. However, when Joan seems to be settling in for a long address on the details of helping her mother at the toilet, telling her repeatedly what day it is and where she is, and so on, Phoebe finds herself checking her watch, stroking her hair, thinking about something else. Phoebe never consciously sends a message, and Joan may have received one only subconsciously, but Phoebe gets a reprieve from the dissertation.

9

In this example, Phoebe is not consciously sending messages of boredom; she would consider that rude. However, these subtle cues do come through, and Phoebe indirectly tells Joan what she is and is not willing to listen to. In this way, the listener leads the speaker. This can happen in a helping situation as well. Perhaps the interviewer thinks a topic is unimportant or personally disturbing or is afraid of addressing a topic that he or she does not feel competent working with. In this situation, Phoebe would do better to identify her feelings and state them directly. Phoebe might say,

> Joan, I understand that these chores are upsetting, however, we have covered this ground several times, and I find myself tuning you out. I feel it might be more productive to look at what doing this means to you; what causes you to focus so persistently on routine and undesirable tasks.

By directly telling Joan how she reacts, Phoebe brings the situation into the open where it can be discussed more productively.

Let's examine a second situation. Phoebe is at a party, where she meets Mario, whom she has known for years.

MARIO: Hello Phoebe. You know, I was just telling Arnie about that time I went to Colorado. I couldn't believe the flight!

PHOEBE: Oh, I know what you mean. That's why I've started taking flying lessons. Now I'll be able to fly myself to any of these places. It was so scary the first time I took control of the plane. . . .

At this point, Mario realizes that the topic is no longer his Colorado experience. Phoebe took a related issue—flying—and brought the topic around to something that interests her, regardless of how the conversation started.

Is this a pattern you've seen before? If you have lost the control of a topic in this type of situation, how did the experience feel? What were you thinking about while the other person was talking? What was the eventual outcome? How did you subsequently approach this person, or did you at all?

Perhaps you were the one who took control. You may not have noticed! If you did notice or were told about it, do you recall what prompted you to change the topic? Do you recall how the original speaker reacted? How might you alter a conversation without using this tactic?

One area in which people may expect to find hidden agendas is sales transactions. Part of the salesperson's job is to guide the customer to a purchase. Customers often fear or detest dealing with salespeople because of the "sly huckster" and "master of pressure and manipulation" stereotypes often attributed to fly-by-night salespeople. Certainly, the economy would collapse without a sales force representing, offering, and moving merchandise and services. The most successful sales forces recognize that the single pressure sale is not ultimately as profitable as a well-established relationship in which customers believe that their needs, desires, and concerns are heard, respected, and attended to. Active listening, within ethical bounds, will help build the trust necessary for long-term sales relationships—and repeat business.

Imbalance of Power.   The imbalance of power is an ethical concern related to the hidden agenda. Power can be attained through physical strength, influence, intimidation, wealth, wisdom, or charisma, as well as through the easily identifiable label of parent, employer, therapist, rabbi, or police officer. Even when a difference in power is only perceived by the people involved, the more powerful person may exert undue influence or the less powerful one may feel coerced or manipulated.

A typical situation involves a supervisor and an employee. Although the supervisor makes a comment meaning nothing more than to suggest a possibility, the employee may feel coercion simply because it is the supervisor who is speaking. An imbalance of power between a client and a counselor could be perceived by a client. The client may see the counselor as someone with mystical powers who can read clients' inner thoughts. There may also be other qualities attributed to the counselor that intimidate the client.

Dealing with the imbalance of power is the duty of the person with the most power. To help the speaker feel open to communicating and to create trust, the listener must ensure that the speaker feels no intimidation or threat. The best way to deal with this imbalance of power is to address it directly as soon as there are hints that it might be getting in the way of open communication. It is neither necessary nor accurate to deny differences in power, authority, or responsibility. A skilled person in a position of power can set the stage for open dialogue by acknowledging these factors and working within the social environment in which they exist.

Training.   Are you as a helper trained to perform the service being provided? It is best to know your limits and be able to refer someone to other professional help. Often a supervisor is placed in the position of counseling an employee. Obviously, the supervisor has a goal: to get the employee to be a better employee. However, the supervisor may not have the training to do counseling of the sort the employee needs.

This lack of training also occurs among professional counselors. It is not hard to imagine all the various types of problems and life situations one could be faced with when working in a clinic. Again, it is best to know one's limits and be prepared to make a referral, when appropriate, to someone with more specialized knowledge.

> **But doctors can't pass a street accident,**
> **nor dips an open handbag, coppers can't pass**
> **a door with a broken lock, Jesuits can't pass**
> **a sin in the making, everyone falls**
> **prey to their training.**
>
> LEN DEIGHTON
> *Spy Story*

Breadth of Knowledge. Staying current on a variety of issues, or maintaining a breadth of knowledge, is related to the topic of training. We all see the world through our own filters, and we tend to see what we expect to see or what we look for. Reading, attending lectures, getting training, and going to professional workshops on a range of subjects within and outside your field will help you maintain a broad focus. This breadth of knowledge will help you see what is really there, not just what you heard most recently.

I have encountered professionals and laypersons alike who could be called "one-issue wonders." Whatever the topic, these people see the world through this single perspective. For example, drug addiction counselors whose training and interests are too narrow may see all relationships as a combination of addictive and codependent behaviors, or a business owner may interpret all behavior as either profitable or costly.

Perhaps the greatest lesson I learned—from a psychology professor I had in graduate school—was to stay broad in focus. This professor occasionally discussed his hobby of astronomy with his students, and his broad focus showed us that being familiar with a variety of subjects can provide useful perspective and analogies. This view also offers a greater chance to find common ground with people of diverse backgrounds. So my advice is to occasionally step back and look at the big picture.

Confidentiality. The need to keep personal matters between participants private is an ethical concern that is vital if trust is to be established and maintained. Most of us have had experiences in which someone violated our trust and told someone else something that was said in confidence. The pain and sense of betrayal, not to mention other damage that may occur, demonstrates how essential confidentiality is to both personal and professional relationships. Trust is difficult to build and easy to destroy.

There are often limits of confidentiality based on laws and on policies of an agency or company. It is the responsibility of the helper, supervisor, or other person in power to discuss any limits of confidentiality before there is a need to break a confidence. In a professional helping relationship, laws designed to protect clients and others require that the helper divulge information about a person's plans to do harm to self or others. When the helper is a student or is being supervised, clients must be informed that what occurs in a session or other interaction will be discussed with a supervisor or supervision group. When an interaction is being audiotaped or videotaped, it is necessary to fully disclose exactly who will view the tape, for what purposes, and when the tape will be destroyed. Whatever your role as a listener, there are laws and school or corporate policies that apply to limits of confidentiality. There are also laws that pertain to violations of confidentiality (Corey, 1991; Corey, Corey, & Callanan, 1993).

Other ethical considerations are also important in providing your listening and helping skills, but it is beyond the scope of this chapter to present them all here. The basic concept to keep in mind is that, as a listener and helper, you are a tool to assist speakers in their struggles to meet their own needs. You must respect people's right to dignity and ethical treatment and accept that they are trying to get along in the world as they understand it.

Summary

This chapter presented a preview of what is to follow. This chapter introduced the pyramid model of active communication. Each level of the pyramid was briefly described to provide an idea of the process as a whole. The pyramid provides a map, of sorts, to follow on your way up the skills levels. The levels of the pyramid are process of communication, attitudes for active communication, skills of the body, probes and questions, skills of reflection, speaking to be heard, and applications.

Ethical considerations were also discussed at some length. It is necessary to provide confidentiality in order to build and maintain trust. Active communication is ethically provided to gain an accurate understanding of another person's messages. To do this, one must avoid hidden agendas and be careful not to indirectly lead speakers to discuss or avoid certain topics. It is better to address these issues directly and state one's own responses clearly.

Next, we turn to the process of communication to build the foundation for active communication.

Suggested Readings

For more information on ethical considerations in the helping professions, see:

COREY, G., COREY, M., & CALLANAN, P. (1993). *Issues and Ethics in the Helping Professions,* 4th ed. Pacific Grove, CA: Brooks/Cole.

Process of Communication

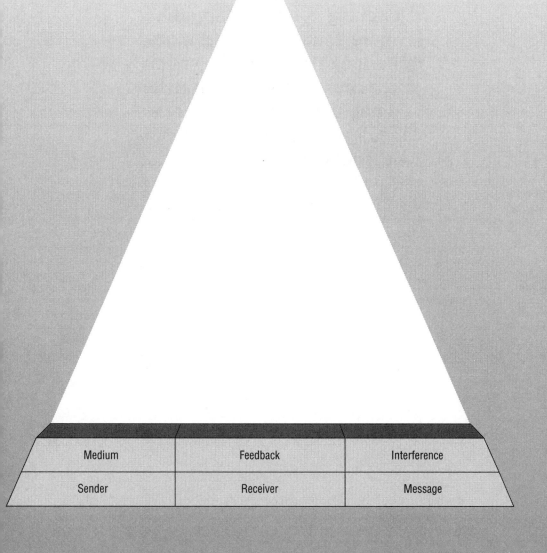

Medium	Feedback	Interference
Sender	Receiver	Message

"I have suggested that listening requires something more than remaining mute while looking attentive—namely, it requires the ability to attend imaginatively to another's language. Actually, in listening, we speak the other's words."

LESLIE H. FARBER

The Ways of the Will

In this chapter, you will be exposed to a brief description of the process of communication. There is more to effective communication than meets the eye. It is a complex system of interdependent processes and objects that must work together if the communicators are to be successful in transmitting their ideas. The purpose of this chapter is to take a close look at how people communicate. This will enable you to communicate knowledgeably, rather than stumbling around hoping that all turns out well.

The concepts of communication presented here include the sender, the receiver, the message, the medium, feedback, and interference. We start by defining the communication process.

What Is Communication?

Communication is the complex process of transferring ideas, information, and images from one mind to another mind. It begins with the intent of one person, the sender, to plant information in the mind of another person, the receiver. From the intent, the sender must translate abstract ideas into symbols, organize the symbols into a sensible message, decide which medium to use, and transmit the message to the receiver. The receiver then perceives the message and translates the symbols into abstract ideas. Finally, the receiver may use feedback to be sure the message is clearly understood. The problem with this whole system is that, at any stage, interference can occur that can change the message. This prevents the sender and receiver from ultimately ending up with the same idea, information, or image.

Why Understand the Process of Communication?

The process of communication is the first level of the pyramid of active communication. It is the foundation upon which everything else rests. When couples are close to breaking up, they often cite the problem as "We don't communicate any more." When companies are having problems in employee relations, customer relations, staff training, or optimum production, the culprit is often considered to be "poor communication." Parents and their children don't just have a generation gap; they also have a communication gap because they don't understand each other's language.

By knowing how communication both works and fails to work, you can make a choice of how to best express your thoughts and overcome

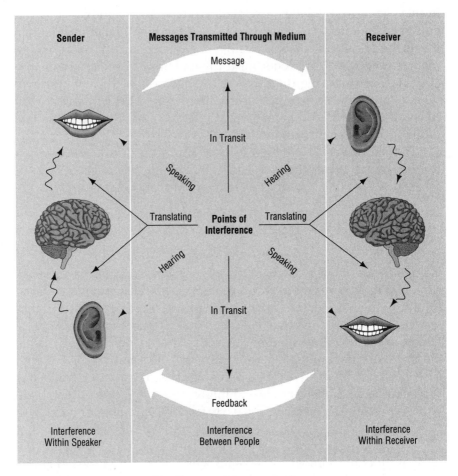

Sender	Messages Transmitted Through Medium	Receiver
	Message	
	In Transit	
	Speaking Hearing	
	Translating Points of Interference Translating	
	Hearing Speaking	
	In Transit	
	Feedback	
Interference Within Speaker	Interference Between People	Interference Within Receiver

The Process of Communication

these problems. This knowledge also makes it easier to recognize that the message you are receiving may not be the message the sender is intending.

In active speaking and listening, it is important to know that communication is not an exact art or science. There is too much variation between human beings for communication to be a simple matter of speak and hear.

Components of Communication

The figure above diagrams the process of communication with each component in place. In this figure, speech is the medium. The thick arrows "message" and "feedback" show a message being sent. The short, wavy arrows show translations of thought (in the brain) into symbols or translations of symbols into thought. The long, straight arrows point to areas of interference, and the labels indicate which process is suffering interference.

SENDER

Senders are the ones who have the abstract thoughts that they want a receiver to understand. Senders translate their thoughts into symbols, then organize the symbols into a message that makes sense to them. This organization may or may not make sense to others.

The sender provides the message for a receiver to interpret. It is the sender's responsibility to select symbols to create a message that the receiver will be able to decode accurately. The sender also chooses a medium that will carry the message. The sender may use a variety of media to portray the same message, such as pictures, charts, and words in a book. Finally, the sender transmits the message, at which point, the receiver hopefully perceives the message.

RECEIVER

Anyone who perceives a message is a *receiver.* This perception can be intentional, such as attending a lecture, or unintentional, as in overhearing a conversation. Either way, the perceiver is a receiver.

Because it is the sender's message, the receiver must concentrate on understanding the underlying intent and meaning of the sender's words and organization. This includes using feedback to learn exactly what a word means to the sender. Technically, while giving feedback, the receiver becomes a sender in a role reversal.

The true challenge of communication is for the receiver to end up with the same final abstract mental understanding that the sender had when creating the message. An active listener is more likely to accomplish this task.

MESSAGE

The *message* is the content—the idea, information, or image— transmitted from sender to receiver. The content is not always the same as the intended message. Attempting to perceive the message in the way the sender intended it to be received is a complex task, because there is usually more to a message than just the words. The tone of voice adds to the message and gives the receiver more information about the sender's intent. For example, the tone of voice used in the statement "Don't touch that" can be begging or threatening. Some other subtle aspects of a message include prior messages that relate to the current one, nonverbal behavior, and emotional content.

It should also be noted that the symbols that make up the message do not duplicate what the sender is thinking. One limit of language is that it uses symbols to describe abstract concepts of the mind. Because symbols are at best a close match to the abstract concept, using a variety of messages and media to send the concept will help ensure that the received message is as close as possible to the intended message.

MEDIUM

The *medium* involves how the message is sent, or by what method of conveyance. Types of media include speech, print or writing, audio, visual, tactile, and a variety of others. We are continually flooded with messages in a variety of media. Simply driving to work, we are inundated with dashboard readouts, traffic lights and signs, lines and arrows painted on the road, street signs, billboards, graffiti, radio ads, songs about desperate love, and our own internal dialogue! No wonder we arrive at work frazzled. Each sender designed a message and employed a certain medium to compete for our attention, because each sender thinks his or hers is the most vital message in the environment.

McLuhan and Fiore (1967) state that "Societies have always been shaped more by the nature of the media by which men communicate than by the content of the communication." According to these authors, one's choice of medium is of great importance to the understanding of the message and to the relationship of the sender and receiver. To illustrate this last point, a telephone call is more intimate than a letter because people can hear each other and get more immediate response and feedback. Face to face talking is more intimate than the phone, and so on.

FEEDBACK

Feedback is any reaction given to what is said and how it is said. It can be a request for clarity or a response to what was asked. With feedback, listening becomes a circle rather than a line. This circular process brings us a great leap closer to active listening.

Active listening exists when the receiver makes a concerted effort to understand the sender's intent, despite the weaknesses in the process of communication. Feedback is vital in this process because it is the means by which one can determine the accuracy of comprehension. In the example that follows, the speaker's thoughts are in parentheses.

> AHMED: (I have to read that book soon.) I need more time to myself.
>
> RAY: (Ahmed wants to be alone now. I should be polite and leave.) Yeah, I know what you mean, I'll see you later.
>
> AHMED: Okay, bye. (I wonder why he took off like that?)

Ray assumed that Ahmed wanted to be left alone but didn't verify that with him, leading to an incorrect interpretation of the intended message. Let's look at how feedback could have circumvented the problem.

> AHMED: (I have to read that book soon.) I need more time to myself.
>
> RAY: (I wonder just what he means by that; maybe he wants me to leave now.) So, do you want me to go now?
>
> AHMED: No, I mean this weekend I have to make some time, not now.

Rather than assuming he knew Ahmed's intent, Ray used feedback to clarify a time frame, part of the message Ahmed was not specific about when sending. Ray had to pay attention to each part of the message and to what types of interference might be encountered at each stage. This way he was able to understand the message as Ahmed intended it.

INTERFERENCE

Anything that disrupts, confuses, or alters the intended message of the sender can be considered *interference*. Interference in communication is more than just "unwanted sound." Because an idea undergoes so many changes while being transferred between people and, in the process, is so unstable, there are many ways the message can be interfered with. Overcoming interference is a goal of active communication.

Interference occurs at every point in the process of communication. We can generalize to two places that interference will occur:

1. Within a Person
   a. When a thought is translated into symbols by the sender.
   b. When symbols are translated into thought by the receiver.
   c. When the person is unable to translate to or from the symbols needed to communicate effectively.
2. Between People
   a. When the message is being carried from sender to receiver.

Probably the most familiar form of interference is that experienced while the message is being carried through a medium: the engine of a passing truck overwhelms the sound of a voice, a smudge on a newspaper hides the print, the moustache scribbled on a face of a billboard distracts from the ad. All these examples of interference show alterations of the messages between the time they are sent and the time they are received.

Another type of interference occurs when a sender does not have the right words or other means to express a thought. If a thought cannot be translated into verbal symbols, the sender must do the best job possible with the vocabulary available. The limitation of a small vocabulary disrupts, confuses, or alters the message intended by the speaker.

Of course, receivers suffer the same kind of interference when they do not understand words the sender uses. (You may have encountered this problem if you have ever traveled internationally.) Even having large vocabularies does not prevent us from having trouble understanding each other's words, because there can be problems with shades of meaning. The same word does not necessarily mean the same thing to any two people. One summer my cousin Hans visited from Holland. When I asked him whether he would like another soda, he replied, "That's o.k." I understood this expression as I would use it, to mean "No thanks, I am fine as I am." I got one for myself and closed the cooler, only later discovering that he had meant, "That's an o.k. idea; I would like another soda."

Fortunately, there is a huge overlap of definitions that we depend on to communicate. Often, a word I use will have a similar definition for both of us, but it won't be an exact match. This is the difference between denotation and connotation. *Denotation* is the dictionary definition, or the standard, specific use of a word. The *connotation*, on the other hand, is the personal meaning a word or phrase has to an individual. There may be no way a person can anticipate what connotations a word will have for another person. Robert Froman suggests that we "don't look for meanings in words but in the person who is using them" (1986, p. 395).

For example, Trisha meets her friend Kenyon while walking her new dog in the park. She invites him to pet and play with the dog, who is straining at the leash to get near him, but he backs away, saying "I'm not much of a dog person."

What is really operating here is that Trisha and Kenyon have different feelings about dogs. Trisha has always had dogs in her family and has always had pleasant experiences with them. When Trisha thinks "dog," the image is of a loyal and friendly companion and protector curled up by the fireplace. In contrast, Kenyon was mauled by a neighbor's dog when he was 7 years old. When he thinks "dog," the image is of sharp, biting teeth; pain; and fear. Until they know each other's connotation for "dog," Trisha won't understand why Kenyon backs away, and Kenyon will wonder why Trisha would want a dog anywhere near her.

A more controversial and highly loaded word that poignantly demonstrates the range of connotations and how far they are from the actual denotation is the word *abortion*. The denotation for *abortion,* in its modern use, is "the intentional termination of a pregnancy." However, with the socio-political/religious battle currently going on, connotations range from "a woman's right to control her body," on the one hand to "murder" on the other.

A speaker may elicit strong reactions simply by using the term *abortion*. By writing this, I may have aroused strong reactions in you because of the connotations this word has for you. You may argue that one of the descriptions of connotations I presented was too strong in one direction. A "pro-life" person and a "pro-choice" person would probably disagree on which description was stated too positively. Consequently, it is the association each person makes to the term that creates the difference of opinion.

The way a person organizes a message can also lead to problems. Receivers are likely to try to interpret the sender's message according to their own ways of organizing. Consider a modern architect describing the idea of a building to someone who has only seen adobe structures; the architect organizes the thought around concepts of steel and glass, but the adobe builder interprets it in terms of mud bricks and poles.

Because of the potential confusion inherent in communication, we distinguish between hearing, listening, and active listening. Active listening requires that the listener look past connotations and other interference. Active communication requires the communicator to listen actively and speak in a manner that reduces interference.

Types of Interpersonal Interactions

There are different standards of communication style in different types of interactions. The purpose and goals of an interpersonal event specify the degree of structure and type of communication appropriate for the interaction. For example, when out to dinner with some friends, we engaged in casual conversation, letting the topic wander at our collective whim without worrying much about covering any topic entirely. After dinner, we entered into a discussion about where to go for dessert, sharing options and opinions on this topic until a conclusion was reached. When we were having dessert, one companion briefly interviewed me about the book I was writing, asking about the topics it covered, when it would be published, how long it was, and so on. She had a specific purpose during the interview, after which we returned to casual conversation.

Although there isn't usually such a fluid switching between types of interaction as in this example, it does illustrate the fundamental characteristics of the three basic types of interpersonal events, or coming together: conversation, discussion, and interview.

CONVERSATION

A *conversation* is a casual exchange of social messages. The goals are generally to enjoy each other's company, to become more familiar with each other, and to exchange ideas. Sharing time together in this manner is designed to be pleasurable in and of itself. There is no external goal—nothing in particular to accomplish.

Conversational interaction is likely to be free-flowing, interruptive, and unmonitored. Some conversation is highly scripted and seems pointless. The "Hi, how are you–Fine how are you?" exchange that occurs in the grocery checkout line seems meaningless, but it helps participants acknowledge each other and sets the tone for the other verbal and monetary exchanges that will take place.

Most conversation varies from idle chit chat to in-depth exchanges people refer to as "solving the world's problems." Topics shift quickly with the end result that participants may stop and wonder, "How did we get on this subject?" Success in a conversation is based more on whether the experience was enjoyed than on whether something larger came of it.

In conversations, roles are poorly defined and can shift among participants quickly and without any formal acknowledgment. Several people may even talk at the same time. Most of us are familiar with the form of cross-talk that occurs, for example, when four people are at dinner and two people continue on one subject while the other two split off to pursue some tangent. Eventually all four return to one topic.

Because the interactions of a conversation are unstructured, an approach to active listening will also be less structured and probably less attentive than in other forms of communication. Although listening skills are always important, fine attention to detail is usually not the purpose of the gathering, and too much attention to form, content, and emotional expression may disrupt the flow instead of serving the participants. Basically, when I'm at a party, I don't engage in a therapeutic analysis of my conversation.

DISCUSSION

A *discussion* is an interpersonal event in which external goals exist. Business meetings are called so that committee members can discuss some topic with the hopes of drawing conclusions, solving problems, or making decisions. Counseling occurs when participants gather to discuss their thoughts and feelings in an open and safe interpersonal environment.

A discussion has a point. When participants wander off topic, the discussion needs to be brought back to the matter at hand. It may be facilitative if the members get along, but this is not a necessary part of meeting the goals of a discussion.

There are often well-defined roles in a discussion. Customarily, someone is designated as the one whose duty it is to conduct the meeting or otherwise maintain the topic of discussion. Another person might be designated to record what is said. Discussion might even be regulated according to formal parliamentary rules, in which speakers must be officially recognized by the discussion leader.

Active listening is a necessary part of discussion. Every member should hear the full measure of others' contributions for there to be any value in having gathered in the first place. Each member should listen attentively, seeking clarification and asking questions as appropriate. Inaccurate listening in a business, medical, educational, or other discussion could lead to disastrous results.

INTERVIEW

An *interview* is a particular type of meeting in which one participant generally seeks information from another through a structured pattern of interaction. Interviews can be thought of as communication with a purpose. Familiar examples are employment interviews and interviews through the TV, radio, or newspaper in which politicians, celebrities, or others are asked to give information about their views, experience, or suggestions.

Another type of interview is the therapeutic interaction. In social work and psychotherapy, an intake interview begins the relationship. The professional seeks particular information from the client to make diagnoses, assess the client's needs, and determine the most beneficial course of interventions. After the intake, interviews may be directed toward unveiling the specifics of the clients' problems and helping them resolve these problems through using their own strengths and applying skills they have learned.

Although interviews differ over the course of a professional relationship, they have consistent features. First, the interviewer must work to establish rapport. One of the essential elements in rapport is the initial greeting and introductions. Ask people what they prefer being called; using an incorrect variation of people's names can offend them. It is usually beneficial to engage in small talk so that people can get a sense of each other.

After some degree of rapport has been achieved, the interview progresses into a focused stage, where what might be considered the "work" of the interview occurs. During the focused stage, participants may work on eliciting information, solving problems, exploring feelings and thoughts, assessing functioning levels and capacity, addressing pertinent issues, negotiating, and other purposive activities (Hepworth & Larsen, 1993; Rogers, 1961).

To effectively close an interview, the participants should summarize what has been covered and agree on what each person will do between this meeting and the next. Finally, participants should address the details of time and place of next contact, any applicable fees, and relevant issues for the foreseeable future (Hepworth & Larsen, 1993).

The hallmark of a skilled interviewer is the ability to listen actively and move the interview successfully through the needed steps. In order to follow the interviewee, the interviewer must be able to ask appropriate and beneficial questions. To structure the interview so that it is helpful or enlightening, the interviewer must have an accurate sense of the speaker's messages.

Summary

This chapter provided a brief background of the process of communication. It was not intended to present all the concepts, only enough to help you recognize what is happening with speaking and listening.

The key concepts presented are interference, sender, receiver, message, medium, and feedback. Interference is anything that disrupts a message from being sent, transferred, or received. The sender is anyone who emits a message, and the receiver is anyone who perceives the message. The message itself is the information being conveyed, whether it accurately reflects the intended information or not. A medium is the mode of conveyance for a message. Feedback occurs when information is received by the sender about, or in response to, the sender's message. This information provides a foundation on which to build the attitudes and skills involved in becoming an active communicator.

There are different types of interactions, each with a different level of structure. Conversation is generally for pleasure and an easy exchange; discussion has specific goals; and an interview is generally a structured interaction in which goals and roles are clearly delineated.

The next chapter, and next level in the pyramid, addresses attitudes for active communication.

Attitudes for Active Communication

Suspending Judgment	Staying with the Speaker	Patience
Empathy	Being Authentic	Postponing Interpretation
Medium	Feedback	Interference
Sender	Receiver	Message

"Our goal is, I presume,
to listen with understanding.
This has to be learned and
practiced. We must become
familiar with this tool, see how
it functions so that we can
make it serve us as
well as possible."

ALFRED BENJAMIN

The Helping Interview

In this chapter, you will read about the attitudes that are consistent with active communication. One of the main points is that attitudes *can* be developed. Holding the attitudes presented in this chapter will help you learn to effectively use the skills presented in chapters that follow.

The attitudes for active communication are having empathy, being authentic, postponing interpretation, suspending judgment, staying with the speaker, and having patience. Each of these attitudes plays a role in helping the listener and speaker communicate effectively.

Discussion of Attitudes

Attitudes are collections of thoughts and beliefs that generally are consistent with each other because they build on a person's philosophy of human nature. Attitudes reflect and guide people's thoughts and actions.

The attitudes presented here create a set of beliefs to guide the efforts that most adequately assist the accurate flow of communication. They build on an understanding of the process of communication in that they primarily reduce one form of interference—that involved when the receiver or listener is not able to accurately understand the sender's message. Perhaps the receiver does not know how to use feedback for clarification, or perhaps the listener is trying so hard to identify with the speaker that the main point of what the speaker is trying to say gets lost. For these reasons, the attitudes for active communication provide the blocks of the pyramid for the level just above the process of communication.

When talking with others, it is important to be aware of our own value structure. The point of listening is to understand the world of the speaker from the speaker's point of view. To do this effectively, listeners need to be aware of their own values, prejudices, anxieties, expectations, and tender spots. To experience you as you are, I must be able to separate what is *me* from what is *you*. As a filter on a camera changes the image that reaches the film, our attitudes can change the image that gets through to our mind.

To enter the subjective world of another person, attempt to understand their connotations. Ask yourself, "Do I know what these words mean to this person? Am I sure?" Through feedback, you can "check out" or clarify the meaning with the speaker. If there is any doubt about what a word or statement means, restate it in a manner that shows you are trying to understand the speaker's meaning.

Let's take another look at the feedback example from Chapter 2.

AHMED: I need more time to myself.
   RAY: Are you saying you need to go now?

When Ray asks "Are you saying," he is telling Ahmed that he is unclear about the precise meaning or intent of Ahmed's statement. The second part of Ray's question—"you need to go now?"—is a restatement of what Ray interpreted Ahmed's intent to be. This gives Ahmed the opportunity to clarify what he meant instead of Ray assuming what time frame he meant.

Thus, active communication requires that both speaker and listener be willing to search for clarification. The ability to do this is built on the following attitudes:

1. Having empathy
2. Being authentic
3. Postponing interpretations
4. Suspending judgment
5. Staying with the speaker
6. Having patience

It may well be a natural part of being human or just part of our culture to act differently from what these attitudes suggest. The challenge is to learn to alter these processes enough to get a clear picture of just what is being said.

Attitudes Can Be Developed

Some people believe that attitudes are developed in childhood and are set for life. This may be the case if the attitude is never challenged or consciously worked on and developed. Certainly, there are consistent and lifelong threads leading through peoples' attitudes, but attitudes can be developed within these bounds. You may already hold some of the attitudes of active communication presented below and just need a way to strengthen them or explore them further.

Probably the most significant factor in developing an attitude is commitment. Attitude change does not occur as a part-time venture. It would be very difficult to hold one set of attitudes at work, and then operate by a different set at night and weekends. There may be a set of behaviors for or an attitude toward work, but the attitudes that make up your personality carry over from situation to situation.

The Attitudes of Active Communication

In this section, each of the attitudes will be discussed in turn. Pay close attention and consider how closely you already match these attitudes.

EMPATHY

In his book *On Becoming a Person,* Carl Rogers (1961) defines *empathy:* "To sense the client's private world as if it were your own but without ever losing the 'as if' quality—this is empathy" (pg. 284). It is my belief that to develop empathy for others requires experiencing one's own feelings. Unless you have experienced sadness, it would be hard to enter the subjective world of a person who is feeling sad.

However, experiencing the emotions that another is experiencing does not require living through the same set of circumstances. I have often heard the argument, "How can you know what I'm feeling when you have never been through this?" The validity of this argument is limited to whether the listener has experienced the emotions, not the situation.

As an example, only one person can know the feelings of being the first person to set foot on the moon. Does this mean that no other person can empathize with Neil Armstrong? We will probably never know his exact feelings, which is what makes each person unique and sets Armstrong apart for his achievement, but surely everyone has felt the emotions of anticipation, anxiety, fear, excitement, and elation, among others, that Armstrong was feeling at the time. I have never been first to step onto the moon, but I have done scary and exciting things that had not been done by anyone I know. So I can say that I have some degree of accuracy in my empathy with Neil Armstrong.

On the other hand, having experienced similar emotions does not mean that you automatically have an accurate sense of what the speaker is describing. Accurate empathic responding requires a careful exploration that helps a listener understand what the emotional experience was for the speaker. It is also inaccurate to assume that similar events provide the same emotional experiences for different people. No two people and no two events are exactly the same.

Empathy can be developed with time and effort by attending to your own feelings and thoughts, observing others' emotional displays, and verifying your interpretation of those displays. Being empathic also requires holding the other attitudes of active communication, because it is difficult to separate your feelings from those of another person.

Your own life and emotions provide a bounty of experienced feelings. In empathy, you draw on those most likely to match another's feelings. The communicator's task is to make sure there is as accurate a match as possible.

A problem often encountered in developing this attitude is that of *inaccurate empathy*, illustrated in the following example.

YUMIKO: I just got my first B grade since I started college. There goes the perfect record.

RAOUL: Gee, that's too bad, I know just how you feel. I got my first B last semester and I was so let down.

YUMIKO: Actually, I'm relieved because now I don't have to be so neurotic about straight A's and I can relax a little. It's like a weight has been lifted.

Raoul had encountered an event similar to Yumiko's—the first B grade—but Raoul had a different set of feelings about it because the experience meant something different to him. He did not explore to see what Yumiko's response was to her receiving this grade, assuming too quickly that her feelings were the same as his because some of the circumstances were the same. Raoul could have striven for accurate empathy if instead he had responded, "I was let down when I got my first B. How about you?"

With this statement, Raoul lets Yumiko know his primary response without assuming her's to be the same, and then he asks her about her reaction. Now Raoul can either recognize the difference and still empathize or express his confusion about Yumiko's response. In this manner, Raoul creates a closeness by letting Yumiko express her response without moving too quickly to try to identify with her and express his empathy. Yumiko is less likely to feel closed off and discounted. As Corey (1995) states, "Empathy dissolves alienation, for the person who receives empathy feels connected with others. Furthermore, those who receive empathy learn that they are valued, cared for, and accepted as they are" (p. 272).

Raoul took what was probably the most important step in achieving empathy with Yumiko when he made a conscious effort to understand her from her viewpoint. Because he took the time to explore her response to the event, Raoul placed understanding Yumiko above trying to tell her what he had felt, yet he still had the opportunity to share his reactions.

Be cautious to avoid the comment, "I know just how you feel." Few statements will close someone down and create interpersonal distance faster. People generally react to this statement as if it were a neon sign proclaiming, "I don't know how you feel, I don't want to know how you feel, and I'm going to change the topic to a story about myself." Possibly the best way to develop accurate empathy is to avoid identifying with, or feeling the same as, the speaker and try to understand that person's feelings. Feel with them, but never forget that you are a separate individual.

Empathy is a necessary attitude for learning the skills that are presented later. The result will be an ability to enter the world of the other person and work at understanding how an experience is for that individual.

🔺 *Exercise — Write about It*

To have empathy, we need to be familiar with our own feelings and with the context in which they occur. Take a few minutes to write about at least three of the following topics. Write about your emotional reactions and your thoughts, rather than making an attempt to summarize the event.

1. The first time you drove with no one else in the car.
2. Visiting a loved one in a convalescent hospital.
3. A song that gets you energized.
4. A song that gets you sad.
5. A time you were betrayed by a childhood friend.
6. A time you were on top of the world.
7. A mosquito bite.
8. Getting a new pet and/or losing a pet.

When you have finished writing, talk with a classmate or other person about each of these topics. You know what each topic means to you; can you explore what each means to the other person? Look for ways that you allow your experiences to color your idea of what they will say. Pay particular attention to how you get past your own experience and resulting expectations to become aware of the other person's experience. Talk about where, when, and how you have had these same thoughts and feelings, what this experience was like, and how it relates to this other person's experience. 🔺

There are, of course, times when someone is speaking about an experience or line of thinking that involves an emotional experience or an intensity of feeling you have not experienced before. For example, many people feel suicidal at some time during life, others report that they have never even considered suicide. This could introduce a scenario in which the listener has little of real value to draw on.

In such a situation, it is best to be honest and direct with the speaker about your inability to respond with accurate empathy. It is most important to acknowledge that this does not make the speaker's feelings wrong or inaccurate for the situation. The rest of this chapter addresses the attitudes that will help you be receptive to this person's experience. Later chapters cover ways to achieve a clearer understanding of what the speaker is thinking and feeling.

BEING AUTHENTIC

> **We listen to every voice and to everybody but not ourselves. We are constantly exposed to the noise of opinions and ideas hammering at us from everywhere: motion pictures, newspapers, radio, idle chatter. If we had planned intentionally to prevent ourselves from ever listening to ourselves, we could have done no better.**
>
> ERICH FROMM
> *Man for Himself*

By definition, to be *authentic* is to be real, up front, genuine, honest about one's reactions, and congruent—to live from the inside and have one's external expressions match one's internal thoughts and feelings. In this book you will frequently read that an attitude cannot be faked; that is what being authentic is about. Authenticity is also vital in promoting and maintaining trust, the key to keeping the lines of communication open.

When people try to mask their true feelings, thoughts, or reactions, they are being inauthentic, and other people in the conversation will probably pick up on this masking. Shakespeare refers to inauthenticity in *Hamlet* in the line "The lady protests too much, methinks" (Shakespeare, 1971, p. 963, Act III, Scene II). By making so much of a show out of trying to convince her listener, the lady indirectly tells that her outer behavior is not consistent with her inner feelings, thus appearing forced and false.

◢ Exercise — Think about It

1. Consider a time when you did not quite trust someone because there was something in that person's behavior or attitude that appeared forced or incongruent with the other things that were going on. What was it that tipped you off?
2. Now, consider a time when you were the one who had to lie or be inauthentic. What was so important about the deceit? Would you do the same today? (The answer may be yes; some people will verbally invade others' privacy if given the opportunity.) Did you think that other people were aware of your attempt at deceit? What in your behavior could have tipped them off? What in their behavior told you that they suspected deceit?   ◢

When people sense that someone is being inauthentic, they may initiate a testing process. This often occurs in inpatient settings and other restrictive environments. For example, James, a high-functioning adolescent who has just entered an inpatient psychiatric unit as "a problem child" is approached by Tran, a case worker. Tran begins by saying, "We are all very happy to be working with you here, and I know I'm glad to be your friend." James immediately wonders why Tran would want to be his friend when they have nothing in common and are on opposite sides of a control structure. He also wonders how Tran can speak for everyone, especially when James has behaved badly toward many members of the staff. James is likely to begin testing Tran by acting out in an inappropriate manner, being defiant, or directly challenging Tran. This process allows James to look for more clues to verify his assumption that Tran's statement was just a line to get James to trust him and to open up.

A more authentic approach might be for Tran to say something that is more reflective of the real situation: "James, I know there have been some problems in the past, but here you have the opportunity to start with a clean slate. You can start new. You appear to be a bright young man. I am looking forward to working with you and hope we get along well."

With this statement, Tran has not spoken for anyone else. He has not been presumptuous about a friendship with someone who has been hospitalized for problem behavior, and he says "get along well" to describe a hope for the future, not a present situation or friendship. Consequently, James does not have nearly as much ammunition with which to fight Tran.

This situation would not have to be modified much to fit a supervisor/employee relationship. Again, because there is an imbalance of power, testing for trust is likely to occur.

Being authentic means that you allow yourself to be the person you are, to own your responses. It means to be aware of and feel your responses, recognizing them as an important way of getting along in your world. And when your responses have an impact, you act on the response, if only to acknowledge it aloud.

Part of being an authentic person means that you also establish boundaries. Everyone needs privacy, although some people are willing to let others see more of their internal world than others are. It is important to identify the degree of openness you are willing to have with different people in different types of relationships. By telling others what your boundaries are and when those boundaries have been reached, you demonstrate respect for yourself and remove ambiguity from the situation.

There is a skill to keeping track of your own feelings, thoughts, boundaries, and responses while also keeping them from clouding the empathy you have for another person. Being authentic while having empathy tells the communicator of the importance of never losing yourself when striving to understand the speaker. This allows the listener to maintain his or her own identity while trying to experience the speaker's subjective world. We can illustrate this striving for balance using Tran and James as an example.

In a therapy group within the hospital, James admits that he has smuggled some drugs into his room; Tran is constrained by the rules of the hospital to report this to the charge nurse. An authentic way of handling this problem would be for Tran to inform James during the session that he has to break the normal confidentiality and his reasons for doing so. Tran could also express his mixed feelings about breaking confidences. (However, it is best to discuss limits of confidentiality *before* they are crossed.)

Here is another example, this time in an employer/employee situation.

ELAINE: (Employee) I have to tell you something that has been bothering me.

JUNE: (Supervisor) I'm glad you came to me. I hope I can help.

ELAINE: I hate to admit this, but I've been taking things home from work.

JUNE: Do you mean stealing company property?

ELAINE: Yes.

JUNE: (Being authentic) Well, you know I can't let this slide. I also have to say that I've just lost a lot of respect for you, and trust. I am glad that you came to me to get this resolved rather than my having to find out myself, but I still have to suspend you for three days, according to company policy.

Or,

JUNE: (Being inauthentic) Well, I don't think this has to be a big problem, as long as you return everything and don't do it again. Thanks a lot for telling me this. (Elaine is later surprised to receive a suspension notice.)

In the authentic response, June demonstrates her sincere feeling about Elaine, then also makes it plain that trust was reduced and that there is going to be a suspension. In the inauthentic response, June does not confront Elaine with her true feelings and leaves the suspension a surprise. It would be safe to assume that Elaine would no longer trust June and would probably hold a grudge because of the lie rather than the punishment alone. Furthermore, Elaine may tell fellow employees of the deceit, increasing their mistrust of management.

As we can see, being authentic can involve a lot of risk. It means being straightforward without benefit of mask or pretense when the mask might feel much safer. However, one must examine the motivations behind stating truths under the guise of authenticity. The ethical guideline for disclosing a response is that it should be for the benefit of the one being helped, not for the helper. Telling someone that "Your sweetheart is cheating on you" or "I really don't think your job is moral," with the caveat "I'm only being honest" may be passive/aggressive hostility rather than simply a clumsy attempt at open and honest communication.

Developing authenticity is a difficult yet wonderful task. Part of its intrigue is that it requires developing an awareness of one's self. I use the separate term *self* because it more accurately describes that part of us that is our identity—the feeling, thinking, living part—in contrast to our body. It excludes the masks and roles we use to get by in the world. So, the self could be considered the core of the person if all the trappings were stripped away.

In *Man's Search for Himself,* Rollo May (1953) describes the alienation of people from their inner self. It is his contention that the industrialized culture we live in requires us to reject our self to maintain our sanity against the onslaught of technology and concentrated populations. Many of the crises we face can be related to this loss of connection with the self—loss of the ability to recognize what the self is trying to tell the rest of the person.

Western technological culture teaches us to focus on thoughts. Although thinking is a necessary and valid part of the self, it is only a part, along with the body, culture and social context, and emotions. Let's concentrate on examining emotions for a moment.

There is disagreement about how many emotions there are— whether there are a few true emotions with several mixtures—and just what the nature of emotion is (Morris, 1993). I am not sure whether pure emotions exist. My experience and observation is that emotions are generally a mixture of feelings. Here is a selection of the subsets I sometimes have found in my own emotions.

Happiness: relaxation and/or stimulation, pleasure, excitement, anticipation, loving and being loved, accomplishment.

Surprise: shock, not knowing what to do, pleasure or fear, curiosity.

Attraction: curiosity, pleasure, arousal, anticipation, anxiety.

Confusion: fear, anticipation, aloneness, ambiguity.

Sadness or depression: anger, resentment, loss, grief, rejection, disappointment.

Anxiety: fear, anticipation, aloneness, structure of unfair rules, frustration.

Frustration: anger, hostility, aggression, aloneness, ambiguity, restraint.

Guilt: sadness, anger, frustration, exposure, fear, self-contempt, loneliness.

You can alter this list and arrangement of emotional clusters to fit your own experience. In the exercises, you will have an opportunity to use this list to identify your own and others' emotional states.

A word of caution before beginning these exercises: in attempting to label others' emotions, it is important to remember that you have only minimal clues about people's true emotional state. What you are doing is trying to maximize your awareness of these cues. Because you are not able to ask others directly for information about their subjective experience at that moment, it is impossible to be sure your guess is correct. In active listening, it is necessary to verbally check that your guess about the speaker's emotion is correct.

Another thing to keep in mind is that you will be projecting your own emotions onto the people you observe. Because you are recognizing behaviors that seem familiar, you will assume that the other person's internal state is similar to one you experienced in the past. In some ways, your guesses about others' emotions may be telling you more about yourself than about the people you observe.

⚖ *Exercises — Write about It*

I. *Label Your Emotions*—This exercise will help familiarize you with your own emotional reactions. By becoming closely aware of your own emotions, you will be better able to recognize the emotions of others. This will help you develop empathy.

For a few days, stop for a moment at different times and spend a few minutes trying to label your feelings. Keep a small notepad on which to write your responses. Experiment with this exercise at the movies, while reading, when waking up or going to sleep, when eating, while talking with others, and during classes. If you have an alarm on your watch, try setting it and answering the following questions when the alarm sounds.

A. What am I feeling right now?

B. What specifically do I feel when I say I am having this feeling?

C. Where in my body do I feel it?

D. What are the specific feelings behind the general emotion?

II. *Label Others' Emotions*—During the following exercise, remember the precautions given just prior to the exercises. Also, don't invade others' privacy by staring or trying to overhear their conversations.

A. Find a place to sit where you can observe other people for about 20 minutes. A lunchroom or park is ideal because there are lots of people interacting and you will not stand out. Select two or three people and watch the way they interact. Try to label their emotions based on behaviors you observe. Behaviors are the external actions that may provide inexact clues to emotions, so be sure to keep behaviors and feelings separate.

1. Are they active or listless, smiling or shouting, and so on?

2. What does this information tell you about their emotional state?

3. How do you know what they are feeling?

4. Look closely and try to determine specifically what clues you have about their emotions. Is it their choice of words or tone of voice (if you can hear them), activities, facial expression, posture, or some other aspect that lets you look into their world?

B. After you have done this exercise a few times, observe the members of a group of about five people. How does each person fit into the group? What does each appear to be feeling? What behaviors do you base your guess on?

C. Another exercise to help you learn to label your own and others' emotions is to watch TV talk shows such as *Donahue* or *Oprah*. There are several to choose from, so pick a few that address a topic that is emotionally charged for you and for the audience.
First, watch the guests.

1. Do they seem to feel supported or accused? something in between?
2. What in their behavior tips you off to what they feel?
3. When guests tell "their story," do they appear authentic? Does the story seem rehearsed or exaggerated?
4. What emotions are displayed and what emotions are stated? What behaviors go along with these stated emotions, and what behaviors suggest the guest is feeling something else?

Second, watch the host or facilitator.

1. How does the facilitator come across to you?
2. Does the facilitator display emotions? Which ones? How?
3. What in the facilitator's behavior makes him or her appear authentic, interested, informed, unbiased, angry, sympathetic, or unsympathetic?

Third, observe the audience. There always seem to be plenty of people ready to solve the guests' problems in a single sentence; something like, "Your parents raised you wrong." Watch how the audience members project their own emotions, experiences, values, and circumstances onto guests. The audience members often provide a model for what you should *avoid* in active communication!

Fourth, observe your own reactions.

1. What do you think about the topic? based on what? Think in terms of what you associate with this topic.

2. What about the facilitator and guests attracts or repulses you? What aspects of this feeling are thought-based versus emotion-based?

3. If the topic, facilitator, guest, or anything else in the program creates a strong emotional reaction in you, stop for a moment, experience what it feels like, and note how your body is responding to the emotion.

Over time, you should become more adept at identifying what people are feeling by observing their outward signs of emotions.

III. *Likes and Dislikes*—Divide a sheet of paper in two and write "Likes" at the top of one side and "Dislikes" at the top of the other. Spend a little time thinking about objects that you like and dislike. List as many things in each category as you can and rate your entries on a scale of 1 to 10, with the higher number being higher intensity of like or dislike.

Once you have depleted your memory, look around you and place each item in the room into a category and label the intensity. Do you hate that lamp but love the desk? like the picture, but hate the way it is framed? Allow yourself to get a little carried away. Spend 10 to 15 minutes or longer listing and rating objects.

This exercise will help you learn about your values and pleasures. There are no wrong likes, dislikes, or reasons. They are all valid because likes and dislikes are completely subjective and personal. You are the authority on what you like or dislike and why.

After you have finished, review your list. What makes you keep some of the things listed in the Dislike column? I am not advocating that you now throw all these things away, but give thought to what is behind your behavior. Maybe you hate the lamp, but you cannot afford a new one on a student's budget. And with the Likes, maybe you love wood furniture but cannot afford or find room for more. For each item, also think about what aspects of it you like or dislike. What would make that lamp more attractive? What specifically do you like about the desk? If you rated something you like a "10" it probably has very little room for improvement; but if it's a "5," consider what you might do to improve its rating. The same is true for dislikes; perhaps you could make some changes in the items with some redeeming qualities to make your environment more pleasant.

IV. *Goals*—Many people move through life with a vague sense of what they want. Others haven't a clue and drift along wherever the current takes them. Think about what you want out of life for:

> the next five years
> this year
> this week
> today

Write down at least five goals for each time period. Long-term and short-term goals provide a rudder that reflect what you want and help you choose what course to take with your time and your life.

These exercises are designed to help you reestablish the connection to yourself and begin learning empathy. Being authentic means being aware of your self and being able to express your self. By using these exercises, you will have gained a greater conscious awareness of your internal psychological processes. You will have begun developing a stronger base from which to generate authentic empathic responding. This process is ongoing, so keep looking at your experiences as a source of connection with yourself and a bridge to other people.   ◬

POSTPONING INTERPRETATIONS

**Only when we realize that there is no
eternal, unchanging truth or absolute truth
can we arouse in ourselves a sense of
intellectual responsibility.**

HU SHIH

An *interpretation* occurs whenever we assume or anticipate that there is more to something than meets the eye. We may assume from a man's appearance and behavior that he is a thief and anticipate that certain thieflike behavior will follow. As soon as we interpret, we place a familiar label on the person, object, or process we have interpreted. Unless we can postpone accepting our interpretations as correct, the tendency is to respond to the person, object, or process as though it fits the label the way a hand fits a glove. The problem is that, when this occurs, we stop seeking further to test our interpretation; we treat it as fact rather than hypothesis.

I believe that all people make interpretations of everything they notice in their environments. It is a natural tendency to look for any potential dangers. According to this line of thinking, everyone makes interpretations of others as a means of trying to survive in a potentially dangerous world.

However, situations that call for active communication generally don't call for survival-oriented snap decisions. Active communication requires that we first recognize this tendency to make quick interpretations. From there, we can learn to view our interpretations as tenuous and subject to modification; they are only hunches or guesses.

Understanding what someone says—decoding the message—involves inferring that each symbol or word has the same meaning for the speaker as it does for you. If you consider the extent of room for error here, given that no two people have identical experiences, it is amazing that we can communicate with any accuracy at all. And this defines our challenge.

Human language has developed over millennia, and the inference of meaning plays a large part in saving time in our speech. The problem is that, in some situations, the inferences are not correct. Consider the amount of inference in the statement

I'm taller than you.

To eliminate the inference and make a complete statement, the speaker would have to state

I, at this moment, when I am standing on my bare feet and as erect as possible, am taller in measurement from the ground at the base of my feet to the part of my head farthest from my feet than you are tall, at this moment, when you are standing on your feet and as erect as possible in measurement from the ground at the base of your feet to the part of your head farthest from your feet.

Note the amount of information inferred in the first statement and the time saved by this inference. Because this is a common statement, the inferences are patterned and predictable to people within the same culture.

When working to reduce interpretations in our communications, we have to keep fact separate from assumption. We often cloud the distinction between what we *know* and what we *believe, think,* or *feel.* Something that can be known is tangible and can be verified objectively by everyone who encounters it. For example, everyone can agree that someone either is or is not crying because crying is an overt behavior. But is the crying a put on, the result of great sadness, or an expression of great happiness?

As another example, when Carmen says, "Drinking alcohol has never interfered with my life," what do we *know?* Do we know that Carmen's life really has never been adversely affected by drinking? All we know for certain is that Carmen made the statement that alcohol has not interfered with her life. Could she be lying, misremembering, joking with us? Could her notion of "interference" be different than ours? The only fact is that she made this statement. We must make several assumptions to conclude that the statement is true or that Carmen and we have the same ideas about drinking alcohol and what constitutes interference in one's life.

Let's turn to what we assume in conversation. Unless there is something to suggest deceit, we generally assume that people are truthful and know what they are talking about. Hearing Carmen's statement, we can probably assume that she knows her life and is willing to tell us the truth unless there is something to suggest otherwise. Maybe Carmen had to come up with a quick answer, and as she has never been arrested, never missed work from drinking, and never passed out in the afternoon, nothing she thinks of as life interference comes to mind. Perhaps, she is giving an answer she thinks the questioner wants. Maybe Carmen's mother is listening in. If Carmen has been brought to an alcoholism counselor by her family or was observed drinking during lunch break by a supervisor or teacher, this would certainly lead us to assume that her statement is not accurate, even if she honestly believes what she said.

The point in separating fact—what we can know—from assumption—what we need to interpret—is not to create blind cynicism but to learn to listen critically. A critical listener is aware of the difference between fact and assumption. Consider the way audience members of TV talk shows overlap fact and assumption. They have the opportunity to listen for a few minutes to someone describe a complex situation that may have developed over years. From there, some audience members fill in gaps with a variety of assumptions and inferences about the guest's personal life, childhood, morality, and more. By being critical listeners—by postponing our own tendencies to infer, interpret, and assume what we don't really know—we can work with the speaker to reveal the entire and accurate message.

Another type of problem arises when our inferences are not clearly patterned, as when the topic or situation is not so common and predictable as comparing our heights. In the following example, Rosa is unaware that Bob is contemplating suicide. Comments that add context are in parentheses.

> BOB: I can't deal with this any longer.
> ROSA: (Inferring) Yeah, I know what you mean. I feel like dropping out (of school) a lot myself.
> BOB: You too, huh? Maybe I should "drop out" (meaning kill myself).

Clearly, Bob and Rosa did not interpret "dropping out" to mean the same thing, and Bob may use Rosa's statement to support the option of suicide. In Bob's mind, Rosa's statement confirms that things are bad for her too and that she is also considering suicide. But Bob misunderstands Rosa, and Rosa may end up feeling confused and distraught if Bob attempts suicide. Consider what happens when Rosa postpones her interpretation until she can verify it.

> BOB: I can't deal with this any longer.
> ROSA: (Inferring but unsure of accuracy.) What do you mean "can't deal with this?" You mean school, or what?
> BOB: No, everything; I mean, I want to end everything.

This time Rosa did not assume that her inference was necessarily correct, and she checked it out through feedback. Rosa got to the true nature of just how hopeless Bob was feeling. It is not as comfortable or safe for Rosa, but she is now in a better position to help Bob.

SUSPENDING JUDGMENT

> **We cannot change anything unless we accept it. Condemnation does not liberate, it oppresses.**
>
> CARL GUSTAV JUNG
> *Psychological Reflections*

A *judgment* occurs when a value is placed on a person, object, or process. As a courtroom judge may pronounce someone guilty or innocent; people make pronouncements of good or bad, harmful or helpful, pretty or ugly, friendly or aggressive, pleasant or offensive. A judgment is based on a collection of interpretations that are then compared against a set of values. The process builds from inference to interpretation to judgment. Ideas formed at each stage become progressively harder to challenge and more resistant to change.

As with interpreting and inferring, it is natural to make judgments about everything we encounter, including people. We learn to label things according to a stereotype and then judge how positive or negative each thing is against this stereotype. There is survival value to this process, allowing us to predict who is likely to be aggressive and who is likely to be friendly. People are intelligent enough to learn when it is safe to suspend judgment while learning more about an individual before acting on the basis of a stereotype.

A judgment may be made at the instant one encounters a new person, but accepting the judgment as accurate and taking action on it can be postponed. This postponement allows time to learn about the individual to see whether the judgment is correct. By adopting the attitude of *suspending judgment,* the listener allows the speaker to communicate personal experience in an environment of acceptance. It also allows the listener to enter the speaker's subjective world.

Separating the person from his or her actions assists in developing this attitude. I am not my behavior; I choose and create my behavior and am therefore responsible for the behavior and its outcomes, but I am not the behavior. The person calling a crisis hotline is not "a potential suicide," but rather "a person who is troubled to the extent that suicide has become a serious option."

Carl Rogers (1961) discusses the concept of *unconditional positive regard.* By this term, he does not mean that one should view everything a person does as being acceptable; rather, the person is a feeling individual who has responses to situations and has a right to have them. Those responses may irritate or anger someone else, but they are part of an attempt to survive in the world as that individual perceives it. As Corey (1991) states, "Acceptance is the recognition of the client's right to have feelings; it is not the approval of all behavior" (p. 213).

Demonstrating through words and behavior that the speaker will not lose your acceptance just because he or she has certain feelings allows the speaker to discuss these feelings. This is obviously important in counseling. A person in crisis may be thinking very socially unacceptable things and have a need to express them. If the speaker feels accepted unconditionally, he or she will not feel as defensive and will be more willing to communicate.

The same process occurs in business settings. For example, an employee will probably be very hesitant to make any negative statements to the supervisor about the job, company, or management. After all, there is a great imbalance of power here, and conditional acceptance could mean loss of promotion or even loss of a job. This sort of pressure heightens the level of risk perceived by the employee.

It is the responsibility of the supervisor to create a climate of trust in which an employee feels free to disclose negative feelings. In this type of situation, the level of trust must be heightened to compensate for the level of risk felt by employees. The benefit to both employees and management is that open communication can help create an environment conducive to better performance. This both raises employee morale and increases company productivity.

Another point to consider in a management situation is that of separating feelings from behavior. The supervisor may be able to permit free expression of negative and positive feelings about the job, but at the same time making it clear that this is not an acceptance of undesirable behavior on the job. In other words, it may be okay to hate the company and hate the job, but remaining employed does depend on maintaining acceptable performance and behavior.

Suspending judgment cannot be faked. The speaker will be sensitive to anything that hints that the listener is creating conditions of false acceptance. And it only takes being caught in a lie once to destroy trust, especially when there is an imbalance of power in the relationship. This information will also spread rapidly among co-workers in a business or among clients or patients in the helping professions. Consider your own experience with people who have violated your trust or whom you did not trust despite their encouragement. Restricting trust is a healthy sign of self-protection.

A popular cultural norm suggests that it's wrong to judge others, as reflected in the common phrase, "I'm not judging you . . . ". Rest assured, a judgment has occurred. When someone says, "I'm not judging you, but cheating on taxes is wrong," the judgment is plain. The comment that cheating on taxes is wrong is a judgment, evaluating the tax evasion as a bad thing. Furthermore, the word "but" generally negates whatever precedes it.

"I really like you, but . . ."
"You're my friend, but . . ."
"I had the ball in the goal, but . . ."
"I did everything right, but . . ."

It doesn't really matter what follows the "but"; you know it negates the leading phrase.

Be careful not to fall prey to the "I'm not judging you . . ." trap. Acknowledge your judgments to yourself so they are less likely to interfere with empathy. Acknowledge your judgments to others when they need to be revealed to maintain communication, trust, or a relationship. Be sure to identify them as your own and as judgments.

◢ Exercises — Write about It

The following exercises are designed to help you identify your own tendency to judge, to examine some of the judgments you make, and hopefully to become more accepting of this human quality. The payoff will be that you will become more aware of what you judge and how you judge, which will make you more aware of when a judgment is influencing your communication and the way you interact with people.

A. *Identifying Judgments*—Because people judge almost everything they encounter, you can use anything you encounter in this exercise. Write about your judgments, personal stereotypes, and expectations about different types of people, things, and behaviors. Make it personal; don't restrict yourself to subjects on this list.
   1. People who own cats.
   2. People who own dogs.
   3. People who always drive within the legal speed limit.
   4. People who don't want children.
   5. Fast food.
   6. Foreign films.
   7. Martial arts.
   8. New York City.
   9. Gossiping.
   10. Cooking.
   11. Driving.
   12. Sleeping.
   13. Interracial marriages.
   14. People who are wealthy.
   15. Welfare recipients.
   16. People who are obese.
   17. Prostitutes (female or male) and their clients.
   18. People who are very attractive.
   19. People who dress in the clothing styles of other countries (whether they are from there or not).
   20. A man with dirty clothes begging for money on the street.

   You will see your judgments reflected in anything you wrote that anyone might think differently about.
B. *Judging and the Talk Show Circuit*—Turn again to the TV talk shows.

1. Identify the assumptions and evaluations people make about each other. Who likes or dislikes whom? Why?
2. Look at your own judgments of people: their behavior, clothes, faces, hairstyles, and so on.
3. What do you think causes them to be who they are and to do what they do?
4. Why are the guests on the show? What do you think they are getting out of this experience?

Everything you wrote about the preceding items probably involves judgments based on your inferences and values. This is not to say that your judgments are inaccurate, just that they are made with incomplete information and therefore involve some degree of uncertainty. Be sure to look closely at what might be a judgment. Going over your responses with a friend or classmate might help you identify each other's judgments. 　　　　　　　　　△

At this point, I need to clarify that I am in no way supporting the discrimination that results from judgments of race, creed, color, religion, geography, gender, sexual orientation, body attributes, and so on. Perhaps the most important purpose of learning about our own judgments and suspending them is to get past the factors that make people easy to categorize. If we are to be effective communicators, we must accept that we make judgments, then dig past the easy categories and look at individual traits, behaviors, and values.

STAYING WITH THE SPEAKER

Staying with the speaker has to do with hearing what the speaker is saying in the here and now of the discussion. It prevents the listener from anticipating what the speaker is going to say and from dwelling on what the speaker has finished saying. There is a delicate balance between (1) hearing the speaker in the here and now, (2) remembering what the speaker has said in order to provide continuity, and (3) staying in touch with one's self. Keeping track of all this can be a tough job that may seem impossible at first, but it gets easier with practice.

Staying with the speaker is tied to the topic of inference and interpretation. In normal conversation, it may be fine to anticipate what someone is going to say. However, when trying to help another person sort through thoughts and feelings, it is best to let that person set the pace. Anticipating what is going to be said disrupts the person's train of thought, can result in leading the speaker, or can put words in the speaker's mouth. Each of these events reduces the level of trust and confidence a speaker places in the listener and disrupts the helpful character of the relationship.

Staying with the speaker does not mean that it is necessarily incorrect to guide the speaker away from unnecessary storytelling. But this guidance requires an awareness of the reasons for telling the story, where it may be leading, and whether there might be a better way of getting to the desired destination. Getting answers to these issues requires a lot of careful interpretation and judgment, and being wrong can interrupt the open flow of communication.

Interruptions, such as the guidance issue above, generally break a speaker's train of thought. Although an interruption can be productive, it must be well timed and executed with the utmost concern for the speaker. The questions to ask before interrupting are:

"What do I hope to gain by this interruption?"
"Am I bored and do I want to hurry the speaker?"
"Is what I am going to say of more benefit to this person than what he or she is saying?"
"Am I interrupting for my benefit or the speaker's?"

Unless there are good answers to these questions, it is generally best to hear the speaker out.

It is difficult in modern society to learn to stay with the speaker. From childhood, we are taught the advantages of "staying one step ahead of the competition." We learn to win games by developing as elaborate a strategy as possible for predicting the opponents' next move or, better yet, next series of moves. Plan ahead, out-think, set them up, and lead them down a losing path that you designed and manipulated them into following—these are the lessons many of us are taught. They are also the indoctrinations and injunctions we must overcome to succeed at active communication.

From personal experience, I know that at times when someone has been helping me sort things through I have sensed that they were expecting a particular response from me. I lost sight of what I had been feeling and became very conscious of what they intended or expected of me. I felt led rather than listened to. Whether or not this person's interpretation and judgment were correct, the poor timing took us both away from the flow of open communication, which was the means to the end.

Another aspect of timing involves feelings. The expression of feelings may take a long time, especially if there are many layers of thoughts or emotions that the speaker had not allowed himself to experience or express previously. If the feelings or experiences are particularly embarrassing or painful, there may be considerable time before they are disclosed. This brings us to the topic of patience.

PATIENCE

Patience is allowing events to unfold at their own pace; waiting when you want to move on; giving others the time to gather and fully express their thoughts and feelings in their own time.

U.S. culture glorifies speed. Many of us have been immersed in the judgment that faster is always better, that the early bird gets the worm. Advertisements push time-saving products. We may come to think of time management as getting more done in less time, rather than using time effectively. Patience, on the other hand, means slowing oneself, waiting, being still. Patience requires an understanding that some things have their own pace, their own schedule. Spring comes when it comes, babies are born when they're born, people understand when they understand. These things can't be rushed without destroying their "realness" or even the very qualities that make them spectacular. Being patient means stepping outside of the cultural rush and letting some things pass, while spending time focusing on others. Becoming a patient person takes a lot of work—and patience.

Letting a speaker say her piece fully at her own pace can require a great deal of patience. In working with elderly persons and mentally retarded persons, I have sometimes wanted to move on because I was bored with the talk, or others had more pertinent things to say, or I had trouble understanding their speech. But to help these people, I had to look at their communications from their subjective worlds, accepting that what they have to say is of great importance to them. And people respond to patience. When they know that someone is willing to hang in there and let them talk, they soon become more trusting and disclose information about themselves that they had previously kept secret or believed no one cared to hear.

Having the patience to let speakers talk unhurriedly allows them the sense of security that you really are interested. From there, they can trust enough to get to the essence of what concerns them, whether it is a work problem, a personal problem, or something else. When working with people, I try to keep in mind that I have a certain amount of time set aside for them and can allow them to use it in the manner and at the pace they need or choose. This allows me to focus on them instead of on myself or where I plan to be next.

A listener allows complete expression of feelings by staying with the speaker. When bottled up emotions are plentiful, having the patience to let the speaker talk about these feelings helps the speaker vent these emotions. Getting feelings out in the open may reduce the sense of being overwhelmed, and it is often enough to help someone calm down. Even if the speaker is angry at the listener, being heard while venting strong emotions can help reduce the intensity and volatility of a conflict situation.

Sometimes, patience means having a tolerance for silence. During interactions, we may feel uncomfortable if there are long periods in which neither person speaks. Silence is often beneficial by allowing a speaker to collect thoughts, consider what directions might be taken, determine how to phrase or present a message, or simply to give time for incubation.

⚖ *Exercises — Try It*

The following exercises are designed to help you develop a sense of patience. Each may involve periods of boredom and tedium. Accept these periods and resist the urge to move on until either the exercise is completed or it is time to move on, whichever the situation calls for.

Remember that one criterion most people use to judge whether another person is really listening is the patience to not rush to conclusions. Also, allowing others to work through their problems, and reach their own solutions requires patience, but in the long run it fosters independence and autonomy. By practicing patience, you will challenge your internal desire to get things moving and familiarize yourself with the calm that comes with unhurried activity. The desire to skip these or similar activities may show you just how necessary it is to do them.

1. *Nature Watch:* Find a spot where you can watch wild birds, squirrels, or some other local animal that is not tame. In a quiet place, put some food out, just a few feet from yourself. Sit still and watch the animals until one comes close enough to eat the food. (Note: don't encourage them to come to you or eat from your hand. Some may bite or carry parasites with disease.)

If you live near a place where deer or larger animals live, settle in a spot near an animal path. Stay quiet, waiting for the animals to pass. It may take a while, but it can be thrilling to see them.

2. *Talk with an Elderly Relative:* You may have a grandparent or other relative who tells stories. Everyone has stories to tell, but some people may take some coaxing. Ask about favorite memories, their wedding, their parents, your parents as kids, school days, toys and leisure, kids' work, and so on. Let them set the pace. You might set

up the time by specifying that you have 1 hour or 90 minutes to spend and would like to hear them reminisce.

3. *Count to 1000:* Count to 1000 but sit and do nothing else while counting. How long does it take? Do you find yourself changing the pace?

4. *Walk Somewhere:* Rather than drive, take the time to walk to school, the store, church, or a friend's house. Walking is a lost art, and it provides a good opportunity to experience the effects of the hurried culture.

5. *Read to a Child:* When you read aloud to children about 3 to 6 years old, they will interrupt, ask relevant and irrelevant questions, point to pictures, and generally slow the process of reading. Make note of your frustrations, how you want the story rather than the audience to set the pace, and how the child is trying to understand.

6. *Meditate:* However the meditation is carried out, it has been said that sitting, focusing on your breathing, and breaking from your awareness of time gives you a sense of freedom, relaxation, and focus that is carried throughout the rest of the day. Sit, facing a wall, in a relaxed posture and focus on your breathing, a candle flame, a metronome, or some other repetitious sound or image. If you feel yourself getting distracted or thinking about something that needs to be done, recognize these thoughts as symptoms of the hurried culture trying to draw you back. They will wait a few minutes. When you've finished meditating, gradually ease back into activity.

7. *Arts and Crafts:* Artistic creations take as long as they take. Rushing art ruins art. Try your hand at drawing, painting, crocheting, doing needlepoint, drawing with colored chalk on your patio, or some other creative venture. When you feel frustrated and want to quit, persist a while longer.

8. *Cook from Scratch:* Whether you "can" or "can't cook" (note the judgment), try creating a meal, dessert, or other culinary creation from scratch. Find a recipe that sounds intriguing, list and purchase the ingredients, and put them together as the recipe directs.　△

Combining the Attitudes into a Whole

You have probably noticed that there is a lot of overlap among the attitudes. They are all part of what it takes to be an active communicator, and they all interrelate and build on each other. It is hard to suspend a judgment after making quick interpretations of what a person's words mean, and it is almost impossible to have empathy when making judgments. It may be helpful to think of these attitudes, as well as the pyramid in general, as a unified whole.

The following statements demonstrate the way these attitudes combine. To have empathy, I need to be authentically interested in the other person. That requires that I don't judge the person too quickly. Because judgments are made from interpretations, I should be careful not to interpret too quickly, and I should stay with the speaker so I am accurate when I do make an interpretation or judgment. This takes a lot of patience.

In the Gestalt tradition, these attitudes are components of this level of the pyramid. But the level is more than just the sum of these attitudes. They combine to form a set that provides a way of looking at people and perhaps a way of life. This level in turn, is part of the pyramid of active communication, which is more than just the sum of its levels.

Summary

Attitudes are collections of thoughts and beliefs that build on a person's philosophy of human nature and can be developed or modified. Because attitudes guide people's behavior and using the skills of active communication is one type of behavior, it is important to examine our attitudes.

Six attitudes for active communication were presented in this chapter. Empathy involves working to understand another person's feelings and experiences as perceived by that person. Being authentic occurs when a person's outward behavior and communication is congruent with internal thoughts and feelings. Postponing interpretation involves setting aside our tendency to think about deeper meanings or themes until we can gain a more accurate picture. Suspending judgment means that we will refrain from evaluating something as good or bad in an attempt to understand another person. Staying with the speaker keeps a listener operating at a pace and within the context of the information and feelings of the speaker. Patience—which had to wait to be discussed last—allows a listener to allow events to unfold in their own time. These attitudes all combine into a unified whole that is more than just a collection of separate attitudes.

The pyramid's next level, which builds on the process of communication and the attitudes of active communication, is the skills of the body.

Skills of the Body

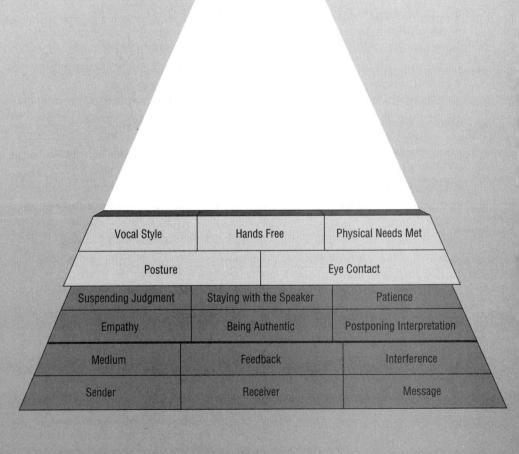

"The ability to be aware of one's body has a great importance all through life. It is a curious fact that most adults have so lost physical awareness that they are unable to tell how their leg feels if you should ask them, or their ankle, or their middle finger or any other part of their body. In our society the awareness of the different parts of the body is generally limited to some borderline schizophrenics, and other sophisticated people who have come under the influence of yoga or other Eastern exercises."

ROLLO MAY

Man's Search for Himself

In this chapter, we will take a look at what to do with your body when you are engaged in active communicating. The body's role in communication has been given increasing attention since E. T. Hall published *The Silent Language* in 1959. It was further popularized by Julius Fast's *Body Language* (1970).

The purpose of this chapter is to help you develop the skills to be used when actually communicating. The first skills to be learned are the skills of the body. Each of these skills relates to silent communication methods that either help put other people at ease or keep them tense.

The skills of the body make up the third level in the pyramid of active communication. These skills are a demonstration of a desire to effectively receive the sender's message, and so they build on the attitudes of the previous level. Furthermore, these skills come before the skills of reflection because it is helpful to first have the body situated and working in a manner that is comfortable and not distracting. This communicates to the speaker that there is interest in the discussion. The following sections describe the skills and how to incorporate them into your communication style. The skills of the body that help you look and feel attentive are posture, eye contact, vocal style, free hands, and physical needs met. It is important to make exercising these skills a habit.

Tuning Your Body for Attention

We begin with discussing each skill separately, delving into the behavioral aspects of performing each skill. Although this approach may seem forced and inauthentic at first, it provides specific practice that will help make performing the skill a natural part of your body language.

Something needs to be said here about coming to a situation prepared. Being prepared for situations that require a high level of attention to someone else, as in a counseling session, requires alertness and freedom from undue concern with other matters. There are different aspects to freeing oneself up. One thing that can be maintained on a regular basis is one's health. Maintaining good nutrition, exercise, and sleep habits are important, because a body that is kept in tune will make fewer demands for attention (through hunger, fatigue, and so on).

Freeing oneself mentally is particularly valuable, and different people have different ways of accomplishing this. I like to spend a few moments getting centered by consciously asking myself the following questions.

1. How does my body feel? Tired, excited, hungry, sore?
2. What mood am I in? Angry, happy, depressed, eager, bland?
3. What is on my mind?
4. What might get in the way of attending to this other person?
5. How do I feel toward this person (or these people) right now?
6. Do I need to eat or use the rest room before meeting with them?

I also make a written note of anything I need to do afterward that I am afraid I might forget. This way, I have a permanent record and don't have to expend energy making sure I remember it while I should be listening.

Other people have different methods of getting centered. Some use meditation, exercise, reviewing of notes, or listing of ideas. Experiment with different things and find out what works best for you. Centering will help reduce the amount of internal distractions, and any reduction of distractions is helpful. While keeping these things in mind, it is also necessary to recognize this as an ideal to strive for but one that is not always possible to achieve.

The Skills of the Body

The skills of the body involve physical activities—things that require more habit development than thought. Before describing each of the skills, it is important to mention that different cultures have different norms for these physical behaviors. For example, in some cultures, it is offensive to make eye contact with the head of a household if you have less stature. Having less stature could mean being younger or even being a female (not all cultures believe in abolishing sexism). Enculturated North Americans often sit with their legs crossed with little concern for just how this is done. In the Middle East, however, it is offensive to sit with the soles of your shoes or feet showing; and while in Thailand, one should take care that the foot not be pointed at anything, as this is considered rude or offensive (Axtell, 1993).

It would be wise for you to learn about the norms of any culture with which you are working. In Southern California, for example, one should learn the norms for Latino and Vietnamese/Indochinese cultures to work in many business, education, medical, and counseling positions. In the Boston area, there are strong African American, Greek, and Portuguese cultural histories and influences. People working in Florida will find it useful to be familiar with Cuban and other Latino cultural norms. Many colleges have organizations and clubs built around cultural or ethnic identities, and they can provide information about their respective cultures. Local Chambers of Commerce may also be helpful in directing you to business organizations geared to minorities. At the end of this chapter, some suggested readings are offered that may help you learn a bit about intercultural issues in communication.

 Exercise — Talk about It

Contact a friend from a culture, racial or ethnic identity, age group, gender, sexual orientation, or other group that you identify as different from yours. Talk with the person about some of the norms of that group that he or she has experienced as being different from your group. You might also want to discuss some of the communication norms you grew up with and get feedback from this person on how they compare to those of his or her group.

Some items to discuss might include:

1. How do you show respect toward parents, toward other elders?
2. When you knew you were in trouble as a child, what was the proper posture and way of responding?
3. What words and gestures are used to refer to something that is liked, disliked, or offensive or to refer to people who are rude, attractive, intelligent or unintelligent, insiders or outsiders?
4. How do you offer a compliment?
5. How do you offer an insult?
6. What are considered appropriate gifts and what things should never be given?
7. What is it like to be from your group and try to communicate or get along with other groups?

Be creative in your discussion, asking about things that strike your interest. As a twist on this exercise, discuss these things with other people you might identify as similar to yourself and see where different norms operate within the same group but in different families.

If you are able to explore several cultures and groups and more than one individual from each, you will better understand that there isn't one right way to communicate, but rather several right ways, depending on who is present.

POSTURE

How you hold your body communicates several things to people around you. A person's mood, energy level, and, in many ways, thoughts are expressed through body language. Consider times as a child when you knew what to expect from a parent based on the posture that parent held when talking to you. Did you know you were in trouble or about to receive a reward? How did you know? Even now as an adult you can key into someone's mood just by how they comport themselves (Fast, 1970).

An attentive *posture* is one that is open to the speaker. Legs that are crossed toward the speaker or are uncrossed, arms that set comfortably at one's sides or with the hands clasped, a body that is turned toward the speaker, hands that are away from the face and mouth—each of these positions is a sign of interest in the speaker and in what is being said. When body positions opposite to these are maintained, the body communicates defense or disinterest. This defensiveness will be noticed by the speaker and will tend to close off communication. Therefore, pay attention to keeping your body positions open toward the speaker in a receptive posture.

Attentive posture may be slightly forward leaning, which is seen naturally in people who are listening attentively. It does not necessarily mean that the whole body is leaning forward. Essentially, a listener brings the head, with its receptive eyes and ears, closer to the source of the message, the speaker. This action is consistent with other observations that the human body opens itself to increased sensory input when interested.

Generally, when deeply involved in a conversation or discussion, people lose awareness of their bodies to some extent. Their focus is on speaking and listening. An attentive posture, however, does not mean that body awareness is cast aside. Sitting positions become uncomfortable, the bladder eventually sends its messages, legs go to sleep, and so on. I am not suggesting that these physical cues should be ignored or somehow transcended. Rather, being attentive means you are not *focused* on the body but are able to shift as necessary to keep yourself comfortable, which allows you to keep your focus on the speaker.

EYE CONTACT

The general rule about *eye contact* is to look at the person you are talking to. This does not mean that you should stare; simply maintain eye contact. The eyes communicate a great deal. Consider how drivers look for eye contact when meeting at an intersection, to be sure each sees the other and acknowledges right of way. Even from such great distances, you can tell whether someone's eyes are meeting your own.

In active communicating, eye contact indicates that you are connected and paying attention. But sometimes eye contact is broken. A break in eye contact occurs any time one person's glance shifts away from the other's eyes. There are positive and negative effects of breaking eye contact. On the positive side, breaks prevent eye contact from becoming a stare. Also, shifting focus to the speaker's hands, a tapping foot, or heaving chest are part of noticing nonverbal expression. Although it creates a break in eye contact, being aware of physical cues can be useful when it comes time to give feedback and summarize what has been discussed.

Sometimes eye contact is broken when someone is formulating a thought into words. Many people shift their gaze away from other people when they are thinking, as if their vision is really looking internally. Be sure to wait until the speaker has finished before breaking from attentive listening, as this break in eye contact may suggest that you are ready to speak and the other person should pause. Breaking away to think also prevents you from hearing what is being said, so wait until the speaker finishes a thought before creating your response.

On the negative side, breaks in eye contact can influence a speaker through feedback. For example, when contact is removed, the speaker may assume that you have stopped listening. It is important to communicate that you are listening to all that the speaker needs to say. Be careful about patterns of breaks; perhaps you, the listener, have problems with these topics and hope the speaker will avoid them. It is important to realize that your actions and reactions as a listener can have an influence on what the speaker talks about or avoids.

The section on ethical issues in Chapter 1 discussed the importance of making sure our own issues and agendas are not imposed on the speaker. After all, in a helping situation, the helper is there to attend to the speaker's agenda. You may not always be aware that you are sending cues to dwell on or avoid a particular topic. Centering before the session and staying aware of your own feelings (to be discussed in more detail later) will help you be aware of when your own discomfort might be getting in the way of listening to what the speaker needs to talk about.

The other side of this ethical situation is taking care of yourself. Don't place yourself in a position of undue suffering. If the speaker is discussing something offensive or irrelevant, be direct in addressing your reactions. Making "I" statements will tell the person how you feel and what you think. Following are examples of "I" statements you can use to change from a subject you find offensive or bothersome.

> "I would rather not talk about people who are not here to defend themselves."
> "I think these jokes are offensive and don't want to hear them."
> "I am uncomfortable discussing politics or religion at a professional meeting."

This way, the responsibility is on you for your own responses, and others are likely to get the message.

Continuing to learn more about yourself—your needs, likes and dislikes, anxieties, hopes, fears, and all the other phenomena that drive you—will help you take care of yourself and remain aware of what might trouble you about topics a speaker introduces. Return regularly to the exercises in the ethics section of Chapter 1, and reach for broader and deeper self-understanding.

VOCAL STYLE

Vocal style includes anything about your voice other than the words you choose. Intonation, fluctuation, pitch, speed, volume, and personal characteristics are the factors that affect your vocal style.

Some voices are soothing and pleasant, whereas others are harsh and hard, although pleasantness is often in the ear of the beholder. For example, I used to wonder how some singers ever made successful careers with their gravelly, rough voices. After listening closer, I came to appreciate this vocal style as part of a type of music. There are not many blues songs that could be done justice by the smooth, crooning type of voice.

There may be characteristics of your voice that people find disturbing. If you are not sure how your voice sounds to others, try recording it and listening to yourself, or have someone give feedback.

In active communication, it is important to be aware of speech. Your vocal style provides information about mood, level of interest, energy level, and whether you care. Spend some time noticing people's vocal style and how it changes when they are happy, sad, under pressure, angry, anxious, and so on. It is often possible to tell people's emotional state merely from their vocal style.

▲ *Exercise — Listen Closely to It*

Tune your TV or radio to a station that broadcasts in a language you do not understand. Listen to the sounds, the emotions expressed by the speaker, instead of trying to understand the words. Is the speaker being passionate or aloof? Is the speaker a seductive deejay, or an evangelical minister? What can you pick up from the manner in which the speakers are expressing their words? Think about how helpful interpreting vocal style will be in telephone conversations.   ▲

Vocal style cues are a two-edged sword. You can pick up a great deal from speakers about their emotional and interest levels, and when you speak you give these cues as well. This is why it is important to understand your own vocal style. Do you truly care about the person and what they are saying? Chances are they will pick up this information by how you speak.

There are many things about a voice that cannot be changed; they are part of your unique individuality. However, you might consider trying to polish those qualities that receive consistent feedback and can be changed. These changeable aspects of one's speech include volume, pace, enunciation, a tendency to speak into one's hand, and so on. See whether the change works for you; if not, you can always change back. The main point is that your vocal style should be congruent with, or match, your emotional state. It is important to remember that others will notice when you are not being authentic.

It is normal for vocal style to change to match different situations. I don't speak the same way in front of a class as I do at home or with friends. Each of us has several different voices, but each voice should be authentic for the demands of the situation. Consider how many times you have come away from someone with the feeling that something wasn't quite "right" about the person. Often, the person did not seem to *be* the way they were *presenting* themselves to be. This is incongruence, or insincerity. You may have picked up subtle cues from vocal style that let on to the scheme. It is the person who is congruent, who is the same on the inside as he or she appears on the outside, who can win trust and be believed (Rogers, 1961).

Clues to a possible disinterested vocal style are any dramatic change, such as a drop in tone or a crack of the voice. Any vocal change tells the listener that a change has taken place in the speaker's emotional state. These changes do not necessarily denote loss of interest, and we surely do not advocate adopting a monotone vocal style, but people will become aware of patterns that appear to show disinterest. What does your voice sound like when you become disinterested?

As with eye contact, vocal changes can direct the speaker to change the course of a discussion. For example, if Anne is talking about considering an abortion and her counselor's voice cracks, rises in tone, or becomes harsh, the counselor may be guiding Anne away from this choice by giving cues that these are not acceptable thoughts. On the other hand, if the vocal style becomes smooth, pleasant and calm, these changes might be giving cues that the counselor favors this option and is glad Anne is considering it. There may be nothing intentional or even conscious about these cues; nonetheless, they are sent and received. Their subtlety, coupled with the fact that we are generally unaccustomed to consciously noting nonverbal behavior, makes these cues even more influential and harder to identify than they should be.

◬ *Exercise — Record It*

This exercise requires some risk, but it is potentially well worth the anxiety, and may help you prevent greater risk or problems later.

A. Audiotape yourself having a discussion with a partner. Try to be as relaxed and natural as possible; it may take a few minutes to get past the awareness of the microphone. Take turns being speaker and listener, and select topics that you are familiar with and are comfortable talking about. Record

65

as least 10 minutes of conversation; up to 30 minutes should give plenty to review.

1. Listen to the recording as if it weren't you talking. Try to be objective. What do you like or dislike about the sound of your voice? Are there things that you do that would make you comfortable or uncomfortable as a listener? Try not to be overly critical, and attend to both the positive and negative.

2. Discuss the recording with a partner or some other trusted person. Keep in mind that the other person can give only his or her own opinion. Be careful not to interrupt or defend, just listen to what the person has to say. You can take or leave the feedback; it doesn't obligate you to change.

3. Review the tape and what you discovered about your vocal style with a professional. The instructor of your class would be an obvious choice. You could also discuss it with a counselor, mentoring teacher, or other professional role model.

It is important to remember that any voice or vocal style is but one part of the total package. Vocal style, alone, won't make or break your future. It is just one piece of the whole puzzle. ◬

HANDS FREE

Attention should be undistracted, which includes keeping your *hands free.* You may have seen cartoons of a psychiatrist playing solitary tic-tac-toe on his notepad. Doing things with your hands communicates that you are only partly listening and that you are also partly concentrating on something else. If you master the skill of active communicating, you will be able to recall major points of the conversation without notes because you will have truly heard what was said. You may be required to take notes during some types of interviews. Although it is usually best to keep full concentration on the speaker, with practice, you should be able to take brief notes without disrupting the flow of communication.

Another type of distraction is to fidget with something in your hands. You may or may not be able to do this without paying much attention to it. However, it is likely that some of your attention is being drawn to the object in your hands merely because it is being handled. Having something in your hands also may draw speakers' attention away from the content of the discussion. They may focus briefly on the object and what you are doing with it, thinking that you are not giving full attention to what is being said: "People become distracted when they are bored. I'm boring this person, I had better hurry up or change the subject." The next time you are talking with someone who is handling something while talking, think about your feelings toward the object, and ask yourself whether you think they are giving you their full attention.

If you are accustomed to fidgeting with something, it may be uncomfortable at first to listen with empty hands. However, fidgeting is a habit that can be unlearned, and I think you will find the benefits are worth the adjustment period.

PHYSICAL NEEDS MET

Meeting your *physical needs* means that you should make sure you are not hungry, thirsty, and do not need to use the bathroom when entering a listening situation. During business meetings, classes, counseling sessions, or other scheduled events, these needs can be met at regular intervals to assure that they don't impose their demands at inopportune times.

I have found that putting off any of these physical needs only frustrates me when I am no longer free to jump up and take care of them. When you know you're going to have a long session, it may be best to use the rest room and get some water, whether or not you seem to notice a need for these at the time. When you often need to be on task for up to an hour at a time, you will soon come to know your own schedules for fulfilling the needs for food, water, and rest room.

Other physical needs that deserve mention are the long-term or lifestyle physical needs: exercise, sleep, nutrition, and general health. Exercise keeps the body generally functioning better. Being toned and flexible allows you to tolerate longer periods of sitting still and reduces the muscle soreness and fatigue.

The importance of sleep is obvious; adequate sleep is necessary for the mind and body to operate at their peak. It can be very distracting and upsetting to a speaker to see the listener yawn. Of course, it isn't always possible to get a regular night's sleep. When this occurs, it is best to tell the speaker that you are fatigued and explain the sleepless night. You may simply say, "I didn't get much sleep last night. I'm sorry for being tired, but it couldn't be helped. I'll try to keep up with you and I'll ask you to repeat anything I miss." Explaining your fatigue at the start of the session will help build trust. The speaker may be able to identify with your fatigue, which can create further bonding.

Overall physical well-being and good health benefits active communication because it benefits life in general. When the body is working correctly, it makes fewer demands. There are fewer aches and pains, fewer illnesses, and often a good feeling associated with having a healthy body.

Meeting your physical needs promotes active communication because it removes possible distractions. Removing distractions allows you to focus attention more by choice than by demand. And this allows you the freedom to focus on the speaker.

Making the Skills of the Body a Habit

By continually working with these body skills, they come to seem "natural" when actually they have become a new habit. Keep track of how you behave when communicating with people. Start paying attention to your physical states of hunger, thirst, fatigue, and elimination needs. When do these strike? Is there a predictable pattern?

Spend some time developing the skills of posture, eye contact, and vocal style. When talking, first pay special attention to one skill at a time, and then blend them. Here are some questions to ask yourself while turning the skills of the body into habits.

How is my posture?
Do I appear open to this person?
Do I look and feel comfortable?
Where are my eyes looking?
Do I stare or look at the person while still allowing breaks in contact?
Do I notice what her body is doing, how her posture is?
How does the sound of my voice come across?
Do I sound happy, confused, sad, or so on when this is what I am feeling?
Finally, what is the whole picture I present?
How do my vocal style, eye contact, and posture blend to present a view of myself as someone who is interested in hearing the speaker and competent to do so?

As with forming any other habit, there will be an adjustment period during which time all these skills may seem unnatural and forced.

Accept this temporary disorganization; it should lessen in a few days and pass in a few weeks. You may notice, or others may point out to you, that you have adopted exaggerated "listening behaviors," being extra-attentive, extra-body aware, intent on your vocal tone, and so on. Again, these are normal parts of developing new skills and habits. When you learned to drive a car, you probably paid full attention to driving, whereas now you probably play the radio, talk to passengers, deal with kids, or a number of other things that would have interfered with your ability to drive at first. The same thing is true for learning these skills. You need to be vigilant in monitoring and producing the desired behaviors in order to settle into a comfortable natural pattern later.

◬ Exercise — Watch It

This exercise will help you develop an increased awareness of the skills of the body. By watching others, you may become aware of the importance of your own body in active communicating.

A. Observe people at a coffee shop or other place where people meet to talk. Notice each person's posture. Does it show interest or disinterest? What about people's eye contact? Do they maintain contact or look at other things? What distracts them? Do they notice movements of the other person, or do they look around the room? Do they notice your eyes watching them?

   I do not recommend listening to strangers' vocal styles because you would also hear their words and that intrudes on their privacy. However, you can notice the vocal styles of people you speak with or people you hear on TV or the radio.

   What can you tell about a person by noticing posture, eye contact, vocal style, and whether their hands are free? Of course, it is harder to notice whether their physical needs are met. Perhaps they seem uncomfortable. This may be a clue.

B. Videotape several different types of TV shows—talk shows, sitcoms, daytime dramas, police shows. Just make sure the selections show people as a central theme. The first time you watch the program, turn off the sound and just watch how the people sit, stand, move, use their hands, and facial expressions. Make notes of the following:

1. What you think people are talking about.
2. What their emotional states are.
3. Who feels what toward whom.
4. Who is in control.
5. Who are the leaders and followers.
6. Who is honest, deceitful, excitable, and so on.

After you have completed this exercise, watch the same shows again and see whether the verbal content supports, clarifies, or refutes your evaluations. Look for what specifically caught your attention and what tipped you off or tricked you into making the judgment you made from non-verbal behavior.

Summary

This chapter presented information about what to do with your body when communicating. Using these skills helps you become comfortable and demonstrates your attention to the speaker. The skills presented are posture, eye contact, vocal style, and having your hands free and your physical needs met. Employing these skills helps you tune your body for attention, and they can become habit with practice.

Each of the chapters presented so far has laid the foundation for the next topic, the skills of reflection.

Suggested Readings

AXTELL, R. E. (1993). *Do's and taboos around the world.* New York: Wiley.

COX, T. (1993). *Cultural diversity in organizations: Theory, research and practice.* San Francisco: Berrett-Koehler Publishers.

FAST, J. (1970). *Body Language.* New York: M. Evans.

GUDYKUNST, W. B., & Y. Y. KIM (1992). *Communicating with strangers: An approach to intercultural communication* (2nd ed.). New York: McGraw-Hill.

MORRIS, D. (1977). *Manwatching.* New York: Harry N. Abrams, Inc.

Probes and Questions

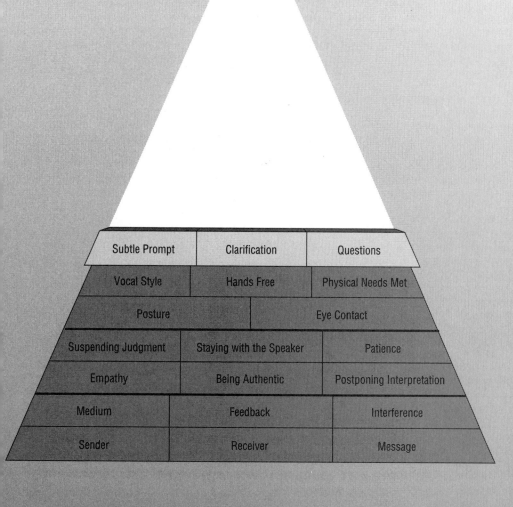

Subtle Prompt	Clarification	Questions
Vocal Style	Hands Free	Physical Needs Met
Posture		Eye Contact
Suspending Judgment	Staying with the Speaker	Patience
Empathy	Being Authentic	Postponing Interpretation
Medium	Feedback	Interference
Sender	Receiver	Message

"The good psychotherapist helps his patient . . . hear his drowned-out inner voices, the weak commands of his own nature."

ABRAHAM MASLOW

The Farther Reaches of Human Nature

In this chapter, we will explore ways of directing speakers to continue speaking, explore a topic with greater depth, or provide information we seek. These probes can be brief statements or gestures that help direct a speaker to move deeper into the content of the discussion. Questions, another form of probe, ask more directly for particular or general information. Questions take greater directoral privilege and can be constructed for more specific responses than other probes can.

As with biology, space exploration, or any other endeavor, a probe is used to penetrate through what is readily present to discover what exists beyond or within. In interpersonal communication, a probe is a verbal or gestural message that explores beneath the surface of what a person is saying. Generally, this type of information gathering encourages people either to continue speaking on a topic or to change subjects. Probes can also be used to clarify and confirm areas of uncertainty.

The most basic and commonly used method for obtaining information is the subtle prompt. We begin our exploration with this method.

Subtle Prompt

The *subtle prompt* is exactly what the term suggests; it is a gesture or vocalization that is just enough to prompt the speaker to continue. Examples of subtle prompts are head nods; comments like "m-hm," "uh-huh," "I see," "I'm listening"; circular waves of the hand in a motion meaning "continue"; and so on.

There is an understood rule in interpersonal communication that one should not monopolize a conversation. The subtle prompt gives permission to continue speaking when the speaker would otherwise fear overstepping the bounds of monopoly. It demonstrates interest without the cumbersome interruption of saying, "Please continue, I am interested in what you were saying." The subtle prompt saves time, and no one's train of thought is derailed.

Another type of subtle prompt involves restating the key word or words in the speaker's statement. The listener uses intonation to rephrase these words as a question by raising the pitch toward the end of the selected word or words. This encourages the speaker to dwell on those words a moment and give more information. Because they are subtle, these prompts create as little interruption of the speaker's line of thought as possible.

For example, let's return to Bob and Rosa and the situation in which Bob talks about "dropping out," meaning "suicide." Originally, Rosa inferred that her notion of "can't deal with" was the same as Bob's. This time, Rosa uses the key word type of subtle prompt to help Bob clarify his meaning.

BOB: I can't deal with this any longer.
ROSA: Can't deal with?
BOB: Yeah, I feel like it's too much, like I just have to get out and disappear.
ROSA: Disappear?
BOB: Yup. Just quit, quit everything, quit life.

Once Bob mentions "quitting life," Rosa knows the level of desperation Bob is feeling. At this point, she can begin questioning him more directly about suicidal ideas. She has not inferred to this point, just tuned into the words in Bob's statements that seemed important or poorly defined. By using this subtle restatement of his words, Rosa prompted Bob to continue a line of thinking he may have been trying to disguise through indirect language.

Exercise — *Prompt It*

A. Observe interviews on TV or videotape. Look specifically for use of subtle prompts. It might be very helpful to watch experienced interviewers and compare them with inexperienced interviewers: for example network newspersons compared to people on local public access cable channels. You might be surprised by what you see in terms of frequency and skill in using subtle prompts. Those who are successful are often good listeners, but not always.
   1. Notice whose prompts are more subtle and whose are more obvious.
   2. Notice the effect the prompts have on the person being interviewed.
   3. What can you incorporate into your own listening style that you see in these interviews?
B. Select two partners from class or two friends to have a discussion with. Take turns being speaker, listener, and observer. The speaker will discuss some topic about which there are personal views but that the speaker has not thought much about. (Use Appendix A for help selecting a topic if nothing comes to mind.) The listener will listen,

using subtle prompts as necessary to direct the speaker and encourage further comment. The observer will attend to the subtle prompts the speaker uses and the effect they have on the listener, being prepared to give feedback to both persons.

Videotape this process if possible, for more accurate feedback. Videotape will also help if there is no third person for an observer. After a few minutes of discussion, the speaker and observer will provide feedback about the following:

1. Number of subtle prompts; too many? too few? just right?
2. Effectiveness of subtle prompts.
3. How the speaker felt affected by the use of subtle prompts.
4. How the listener felt affected by the use of subtle prompts.
5. How the observer felt affected by the use of subtle prompts.
6. What might be done in the future to improve the listener's effective use of subtle prompts.

By paying attention to his or her own use of subtle prompts and using the feedback of others, a listener can learn how to be most effective with this skill. It is not a science of expertise; rather it is a skill born of practice to encourage speakers to continue when they may feel anxious about what is being said.

Clarification

**A good listener tries to understand
thoroughly what the other person is saying.
In the end he may disagree sharply, but
before he disagrees, he wants to know exactly
what it is he is disagreeing with.**

KENNETH A. WELLS
Guide to Good Leadership

Clarification uses requests to obtain more specific and well-defined information. It serves to clear up possible misunderstandings between speaker and listener. In this respect, clarification is also related to paraphrase and summarization, discussed in Chapter 6.

The need for clarification may be felt when you sense that your definition for a word may not be the same as your speaker's. Often, there is a sense that something "just doesn't fit," or a phrase or concept used by the speaker is unclear. At this point, seek clarification by asking, "What do you mean by_____ ?"

Recall from the conversation between Bob and Rosa that at first Rosa inferred that her definition of the phrase "drop out" was the same as Bob's—but she was wrong. However, by asking Bob just what he meant by the phrase, Rosa was able to clarify what Bob meant. She then understood that Bob was considering suicide instead of just quitting school.

Some types of words and phrases are more likely than others to require clarification. Slang and jargon consist of words known primarily to subgroups of a culture. Both are ways of identifying in-group members; slang is intended to confuse outsiders and jargon is used to communicate specialized information to members of an occupational group. (I used to wonder how a wine could be "dry" when it is essentially a liquid.)

Euphemisms are substitutes for words that are unpleasant to say. For example, the phrases "passed on," "left us," "went away," or even "bought the farm" are less direct than "died," but they are less distressing for some people to say. There are countless euphemisms for death, sexuality, status, emotions, and any other topic with taboos against frank discussion. Euphemisms are generally neither better nor worse than the words for which they are substitutes, but they are an inexact way to communicate unless the specific terms and meanings are familiar to both speaker and listener.

The best thing to do when one encounters slang, jargon, and euphemisms is to ask the speaker to stop and define the words in language familiar to both speaker and listener. This is the essence of clarification. Be prepared to hear a lot of slang and jargon from some people. They may spend so much time speaking in these specialized ways that they are not accustomed to speaking with people outside their own reference group.

Much of every "generation gap" is a result of a communication gap. How often have you been involved in a parent/child conversation and not known what all the words meant? This same type of gap exists between professions. Imagine a welder talking with a musician; they work with different concepts and use different jargon. The same is likely to be true between counselor and client, supervisor and employee, and even friends.

It is also important to remember that, as a trained listener, you may be the source of confusion rather than clarity. After learning to talk about "authenticity" and "subtle prompts," it may be hard to recall that not everyone knows what these terms mean. Be aware of which words were unfamiliar to you when you first heard them, and realize that these words will probably be unfamiliar to others.

When having a particularly difficult time clarifying what someone has said, an effective strategy is to request that the speaker "say it as though you were only 5 years old." For some people, it is best to shift the focus onto yourself and request that the speaker "say it as if I'm only 5 years old." It may take a little convincing to loosen the person up to the idea of speaking in a childlike manner, yet it can be very effective because it breaks down all the pretense and careful rehearsing people do when they are trying to say something "right." You may also find it helpful to do this yourself when others appear to be confused by your own statements. Say, "Let me try to simplify what I mean . . ." and present your own message in simpler language.

◬ Exercise — Clarify It

Take turns practicing clarification with a class partner, alternating between being speaker and listener. The speaker should discuss some topic that is unfamiliar to the listener. Clarify this topic before beginning. Choose a hobby, major course of study, travel experience, talent, or something that the listener does not know much about.

The listener will practice clarification skills. Whenever something is unclear, the listener should seek clarification. After a few minutes, stop and give feedback. Both speaker and listener should discuss how the discussion went. Be sure to clarify during the feedback session.

Reverse roles and repeat.   ◬

Questions

"Whowhathowwhenwhere!?"

DAD

A *question* is a request for more information. Skillful use of questions is an important part of active communication. There are several factors to consider in asking questions: goals, type of speaker, type of situation, physical environment, relationship, and phrasing.

Goals. Be mindful of just what it is that you need to know and toward what ends you want the information. Before even considering constructing a question, consider the following points.

Are you looking for specific or general information?
Do you expect a long or a short answer?
Is this question appropriate with this person?
Is this question appropriate at this time?
Is the information necessary for working toward your goals?
How will this information be beneficial?
Are the goals you are working toward going to benefit the respondent?
If you are looking for information to satisfy your own curiosity, have you made this clear to respondents and obtained their informed consent?

Always keep in mind the goals that drive your desire or need for information. When you are asking out of your own curiosity and not to assist the speaker, then you have become a voyeur and have violated the ethics of the helping relationship.

Type of Speaker. Do words gush forth from this person? Does getting information seem like pulling teeth? Do your questions need to provide guidance and help the speaker stay on track?

Sometimes the speaker's personality creates a need for particular skill and care in questioning. A person's culture, gender, history of personal experiences, and other factors may create a particular way of responding. Some speakers are eager to tell what they think and feel. These people are easy to ask questions of, and you may have difficulty limiting their responses. Others will be more guarded or may have a broader sense of privacy.

Type of Situation. The situation could be a counseling session, a supervisory meeting, a parent asking about a child's first date, or any other situation in which information is requested. As you might expect, people will generally be more or less willing to offer information depending on the situation. If people fear that disclosing information may lead to harmful consequences, such as when a family is accused of child maltreatment, they may be guarded about the information they offer. In other circumstances, such as a social gathering in which people are asked about their opinions on social issues, people may be more willing to speak up.

The situation also involves the purpose and goals for the meeting. The questions for a job interview may be highly structured. A social services intake interview will also be structured, but in a different sense and with different goals. Both situations differ from a parent/ teacher conference, in which the information sought will differ considerably. Careful thought into how questions are structured will be necessary for success in each type of situation.

Physical Environment.   The location and arrangement of furniture can influence how much a person discloses. People may be less likely to discuss personal issues in public, whereas the privacy of an office or home may help relax inhibitions.

Even different types of office furniture and their arrangement can impede or assist disclosure. Perhaps the most intimidating and restricting example would be a hard plastic chair sitting across from a high-backed executive chair on the other side of a large desk covered with papers. This setting connotes a major power difference between the two people. Comfortable furniture, arranged face-to-face, creates a more open feeling. If at all possible, have a place for each person to sit in equivalent chairs with no obstructions between them. If a table is necessary for paperwork, try to sit along adjoining sides where a minimum of the table is between. This will maximize the sense of equality and will facilitate more open communication.

Relationship.   Is one person the other's boss? Are the people friends or acquaintances? Do they get along or experience conflict? Is one requesting favors from the other? The simple fact that one person has the authority to ask questions of the other suggests that the asker is being placed in a more powerful position than the respondent. The owner of a company, for example, doesn't have to answer questions about being late for work; employees don't have that privilege.

The relationship is always part of the communication dynamic. The history of the participants' relationship creates a depth to the interactions that goes beyond what is being said or done in the moment. This dynamic is readily apparent when former lovers meet. A great deal of business is conducted in athletic clubs, at lunches, and on golf courses, far from the official negotiations. This reflects an attempt to create a relationship that will ease the more formal transactions.

Perhaps the most important things a questioner can do to facilitate the relationship dynamic is to hold and portray the attitudes for active listening discussed in Chapter 3. Being authentic, building trust by being trustworthy, displaying accurate empathy, and having patience all help create the type of relationship in which communication can flow more easily.

Phrasing. Differences in phrasing produce three basic types of questions: closed questions, open questions, and "why" questions. Sometimes these questions are embedded in statements or otherwise disguised. Question types differ primarily in how much and what type of information they seek.

It takes some practice and skill to be able to phrase questions that will provide the desired type of response and amount of information from a particular speaker in a particular situation. This section will devote a considerable amount of time to developing this skill. By the phrase "desired type of response," I refer to length and depth of expression, not an attempt to direct the person to alter the content of thoughts or feelings. Such alteration would result from leading the person, which is contrary to the goals of active communication.

How to blend these aspects of questioning—and their importance—becomes easier to see after observing how they can affect communication. A single question can evoke different types of responses from different types of people in different situations. Consider the following example, in which Quincy asks questions, Arthur gives long-winded answers, and Alice keeps answers too short.

QUINCY: How did you feel about that?
ARTHUR: Well, first I was angry. I really was put out, I figured that was just rude and thoughtless. But over time I began to see their point of view and it didn't bother me so much . . ."
QUINCY: Alice, how did you feel about that?
ALICE: Angry.

The questioner needs to be careful about wording questions for Arthur because it may be hard to get simple information. Arthur has a lot to say and has no hesitation about taking the opportunity to say it *all.* But if the questioner only needs to know a specific bit, it may be necessary to preface the question with "In a word . . .", then insist that he limit his response to one or two words. Because long-winded Arthur seems to be quite willing to talk, not much care needs to be taken in phrase design to help him open up.

With short-answer Alice, the questioner may have to take a careful look at phrasing. She appears to be closed off to discussing her reaction to the event being discussed. It may be necessary to look into trust issues, boundaries of privacy, or willingness to face her own reactions. Her defense must be respected, because she possibly feels the need to protect herself. If this need is disregarded, it will most likely result in stronger defenses. However, if the situation is one in which she has expressed desire to examine feelings about this issue, it may be appropriate to challenge her to press on. Alice just needs clarification that she should express herself more fully. A statement to this effect would then resolve the problem of overly restricted answers.

In response to Alice—especially if their relationship is well established and strong—the questioner might come back with another question. Here are some examples of responses Quincy could use to probe deeper into what Alice meant by the term *angry*.

1. "Tell me more about the anger."
2. "Specifically, what do you mean when you say you are 'angry'?"
3. "Please describe the anger for me. (Allow her to respond.) Where in your body did you feel it? (Allow her to answer.) What did feeling angry do to the way you felt and reacted?"
4. "What does it mean to you to feel angry?"
5. "Angry?"

The first three examples are more directive than the other two and will be harder for Alice to decline to answer. It can be manipulative on the part of the questioner to introduce these types of questions. Whether these questions are appropriate depends on the permission Alice has granted to the questioner about how far to push in probing for feelings or thoughts.

Question 4 is a bit abstract. Alice may be unfamiliar with looking into the meanings behind feelings or the way her feelings today relate to past situations. Question 5 is just repeating the word with the inflection of a question. You will recall this also fits the definition of a subtle prompt.

If Quincy is her therapist, Alice may have said, "This is important to me. Please challenge me when I try to close feelings off." On the other hand, if Quincy is her supervisor at work, she may be thinking, "My feelings are none of his business."

Thus, although phrasing is important, a certain type of question does not guarantee a certain type of response. Questions must be taken in the context of the speaker, listener, environment, relationship, and other factors that might be at work. Categorizing questions as either "open" or "closed" provides a basic structure and clues to what type of phrasing should work for a particular goal.

CLOSED QUESTIONS

A *closed question* is structured to elicit a limited and specific response and attempts to prevent the respondent from saying anything more than the minimum required response. Closed questions are also generally simpler and more direct than open questions.

The purpose of a closed question is to obtain specific information, such as names, numbers, and locations. Closed questions ask for basic facts about people, about who they are in terms of demographics rather than individual and unique experience. Closed questions are often used when people first encounter each other, whether it's at a therapy session, job interview, or cocktail party.

Here are several examples of closed questions. See whether you can identify how the structure and phrasing attempt to limit the response to each question.

"What is your name?"
"How long have you been in Tennessee?"
"What job did you hold before becoming an artist?"
"Did you ever see *I Love Lucy*?"
"Will you go out with me Friday night?"
"Do you like turtleneck sweaters?"
"Where is Ed?"
"When did you get married?"
"Who is your favorite jockey?"
"What time does the movie start?"

Do you recognize the common elements that make each a closed question? Think about this again, but as you do, go back over the questions and answer them. What is it about your answers that is consistent? Remember, content does not count in this analysis as much as structure and phrasing.

One key to the structure of a closed question is the first word. The closing effect of words such as *what, do, did, who, when, where,* and *how long,* is that they ask for specific information, especially when the specific category follows the first word. Any question that requests a time frame, name, place, date, or identification of something specific is a closed question.

Each of the questions listed could be answered fully with one or two words, the minimum required answer that would satisfy the request for information. It would be hard to give long answers without breaking some of the rules of language used in the questions.

Think of the response in terms of a closed book as opposed to an open one. With an open book, you can read all the content and get to know the book fully. With a closed book, you can only see the title, author, how thick it is, and guess at its age. By using closed questions, you will find out only this type of limited information.

◢ Exercise — Ask Closed Questions about It

A. *Generating Closed Questions:* For the following scenarios, write down some closed questions you would ask to gain further information. Be sure that the questions are phrased to obtain particular and specific information. Brainstorm, writing all the closed questions that come to mind without concern over whether you would necessarily ask each question. (There are more scenarios in Appendix B.)

1. A teenager comes home from his date. The couple has been dating steadily for 2 years. On this evening, they were robbed at gunpoint before entering a restaurant.

2. A woman tells you that, at age 42, her well-planned life is falling apart. Six months ago, she was laid off from the job she had had for 16 years, a week ago she caught her kids smoking marijuana, and today her doctor told her she's pregnant.

3. Your client (or patient, student, or employee) tells you he was diagnosed with cancer. He wants to refuse treatment because he saw his mother and his uncle both die from cancer despite intensive treatment. The doctor said his cancer is not terminal at this early stage.

After you have written the questions, review them to be sure they really are closed questions.

Now, go through the questions and determine those that you would actually want to ask, versus those that would not be very productive. Judge productivity according to what information would be helpful in working with the person.

Examine the questions to see what assumptions led you to think that the questions you chose will elicit valuable information. For example, in scenario 1, did you ask the young man what section of town the couple was in? How does this make a difference in helping the young man?

B. *Role Playing Closed Questions:* Form groups of three, with each person taking a role: questioner, informant, observer. The questioner asks closed questions about the topic described by the informant. If the informant is having difficulty thinking of a topic, refer to Appendix A. The informant begins by giving a little background for the topic, then answers the questioner. The observer records the closed questions and types of responses, then gives feedback after the questioning is finished. Trade roles until each person has had a turn in each role.

Allow about 5 to 7 minutes for each turn.

When phrasing a question, consider that it can invite the respondent to either go into lengthy detail or remain brief. Setting up conditions for length of response is in the wording more than in the intent, as in the example "How did you feel about that?" given earlier. We don't know, for sure, what the intended response length was, but we do know there were very different lengths to the answers given by Arthur and Alice.

There is a difference between predicted and possible responses to questions. *Predicted* responses to the example closed questions given earlier could be: Barbara, 6 months, teacher, yes, no, yes, fishing, 1976, Jenson, 7 P.M. The questions took longer to ask than to answer. But a great deal more could have been said; *possible* answers may be very long and involved.

Sometimes long answers are fine, but when gathering specific data, it may be necessary to state the intended brevity of response directly. A good way of breaking off a long response is to say, "We'll get to all of this in just a moment. Right now I just need the specific answer." This reminds the person that responses should be brief.

An ethical issue of control arises with regard to closed questions. Especially in a therapeutic or counseling situation, the questioner has to remain conscious of whether questions are designed to avoid certain topics or to guide the client. The question to ask is, "Am I putting my own needs before those of my client?" This can also be an issue in the workplace. Many of us have heard the comment, "The boss may have an open door policy, but there's a screen door that no one can get through." That screen can be the types of questions asked and the types of responses allowed.

Sometimes questioners use structure to avoid discussing topics that may bring up their own painful or uncomfortable feelings. If you notice yourself doing this, evaluate your thoughts and feelings about the topic and the speaker.

OPEN QUESTIONS

The *open question,* often referred to as an open-ended question, requests that a speaker give a personal narrative of what is being discussed. It asks for longer, fuller answers and invites the person to speak freely in response. Open questions encourage people to express more of their internal worlds, including thoughts, opinions, evaluations, feelings, memories, hopes, fears, and other reactions, by structuring the question in such a manner that it tells the other person that full narrative, detailed description, or other extensive information is being sought. Open questions seek to guide and direct the speaker to full expression.

Returning to the book metaphor, in which closed questions are like looking at the outside of a book, open questions are like opening the book. They are requests to be allowed to read the "book" of another person. Open questions seek to discover the hidden qualities of "who" a person is that could not be discerned by looking only at the person's demographics.

In the following examples, see whether you can identify the structure and phrasing that help speakers open up and tell more of their experiences.

"What else was going on when the fight broke out?"
"What are you thinking about?"
"What does it mean to you when you say you feel angry?"
"What did you expect to happen after you took the ball from your brother?"
"What do people in your family do around holidays?"
"What plans do you have in case some tragedy happened in your family?"
"What was special about your favorite elementary school teacher?"
"How does having children affect your marriage?"
"How can I help?"
"How does your training qualify you for this job?"
"How are you planning to improve your grades?"

Perhaps the most obvious quality of these questions is that each begins with *what* or *how*. These two terms, more than any others, express the desire to hear expanded detail. Questions that begin with *why* are also generally open but tend to be judgmental. "Why" questions will be discussed later in more detail.

Using open questions gives speakers the freedom to describe a situation as they perceive it. For example, asking the closed question, "Are you married?" offers the respondent two basic options from which to choose: "yes" and "no." This doesn't allow much room for being intimately involved, and the asker may not receive quite the accurate information being sought. "No" would have to be the response of anyone who is cohabiting, is in a gay or lesbian relationship, has never been married, or is divorced or widowed. "Yes" would be given by heterosexual married couples, gays and lesbians who identify their relationship as marriage, and bigamists. But how does someone answer when the spouses are legally separated—neither really still married nor yet divorced? Clearly, it would be easier to ask the open question, "What can you tell me about your close relationships?" (Worthington, 1992).

▲ Exercise — Ask Open Questions about It

This exercise is a duplicate of the one on closed questions. This provides the opportunity for side-by-side comparison of the effectiveness of each type of question in gaining different types of information.

A. *Generating Open Questions:* For the following scenarios, write down some open questions you would ask to gain further information. Be sure that the questions are phrased to encourage full exploration. Brainstorm, writing all the open questions that come to mind without concern over whether you would necessarily ask each question. (There are more scenarios in Appendix B.)

1. A teenager comes home from his date. The couple has been dating steadily for 2 years. On this evening, they were robbed at gunpoint before entering a restaurant.

2. A woman tells you that, at age 42, her well-planned life is falling apart. Six months ago, she was laid off from the job she had had for 16 years, a week ago she caught her kids smoking marijuana, and today her doctor told her she's pregnant.

3. Your client (or patient, student, or employee) tells you he was diagnosed with cancer. He wants to refuse treatment because he saw his mother and his uncle both die from cancer despite intensive treatment. The doctor said his cancer is not terminal at this early stage.

After you have written the questions, review them to be sure they really are open questions.

Now, go through the questions and determine those that you would actually want to ask, versus those that would not be very productive. Judge productivity. Judge productivity according to what information would be helpful in working with the person.

Examine the questions to see what assumptions led you to think that the questions you chose will elicit valuable information. For example, in scenario 1, did you ask what the young man did after the robbery? How does this make a difference in helping the young man?

B. *Role Playing Open Questions:* Form groups of three, with each person taking a role: questioner, informant, observer. The questioner asks open questions about the topic described by the informant. If the informant is having difficulty thinking of a topic, refer to Appendix A. The informant begins by giving a little background for the topic, then answers the questioner. The observer records the open questions and types of responses, then gives feedback after the questioning is finished. Trade roles until each person has had a turn in each role.

Allow about 7 to 10 minutes for each turn.

Of course, minimal responses to even the best phrased open questions often do occur, which can be one of the most frustrating times for the active listener. At times like these, as with short-answer Alice in the example earlier, it is best to take cues from the speaker. What is it that this person is afraid of? Is it trust? Is it the subject? Is there some perceived threat about divulging this information? Or is it something else?

Rather than pressing for openness right away, it is usually best to follow the speaker's lead and change the direction of questioning. Deal with the source of interference, then come back to the topic. When there is resistance, it is generally a sign of some type of fear. Explore the fear. Acknowledge that you feel or sense something is getting in the way or that communication is not occurring. Then ask, "What prevents you from talking about this?" or in some other way discuss your feeling that the speaker is protective or defensive. This helps establish trust by demonstrating that you are sensitive to the speaker's needs and emotions.

EMBEDDED QUESTIONS

Embedded questions are statements that contain a request for information (Hepworth & Larsen, 1993). They are usually responded to as if a question were asked but have the benefit of creating a more conversational tone, so the respondent is less likely to feel inter-rogated. Embedded questions often start with phrases such as "I wonder . . .", "I'm curious about . . .", and "I'd like to know . . .".

Although inclusion of embedded questions breaks up the monot-ony, overuse creates its own monotony. Furthermore, when requests for information come primarily as embedded questions, the asker ap-pears indirect and respondents may distance themselves.

POLITE COMMANDS

A *polite command* (Hepworth & Larsen, 1993) is a statement that has the same effect as a question but *directs* the respondent, rather than asking a question. Although a directive to say more seems demanding, the essential quality of the polite command is that it encourages a speaker to say more through a polite and respectful request.

"Tell me about your feelings for each other."
"Tell me how your behavior gains you respect."
"I'd like you to begin where we left off last time."
"Give me some detail about the car crash."

When using polite commands, be sure to be respectful of the respondent. If you find you have difficulty in getting the responses you expect, be sure to seek feedback about your methods from a classmate, instructor, or other experienced listener.

As with embedded questions, overuse of polite commands disrupts the flow of a conversation or discussion, rather than improving the experience and gathering the information that was sought.

There are some important ethical considerations to keep in mind. The question must be asked, "For whose benefit is this person being asked to divulge the private information?" In a legal interrogation, the speaker's personal growth or problem solving are not at issue, and the therapeutic benefit of building trust is disregarded. However, in a therapeutic, educational, medical, or other counseling situation, the importance is generally placed on the process of building trust and establishing a constructive relationship. In this climate of trust, private matters may be revealed, and it is often the act of revealing, rather than what is revealed, that is important. Similarly, a therapist, coun-selor, or supervisor who places the importance of the information above the person has lost sight of the purpose of active listening and is simply being nosy. Always keep in mind the goals that drive your desire or need for information.

Avoiding Problems with Questions

There are two sides to every question.

PROTAGORAS

A WORD ABOUT "WHY" QUESTIONS

You may have noticed that none of the examples given used a question beginning with the word *Why*. The reason is that, as an active communicator, the questioner is seeking information and understanding. As Virginia Satir (1988) aptly points out, "Generally speaking, 'how' questions lead to information and understanding and 'whys' imply blame and so produce defensiveness" (p. 136).

The defensiveness Satir discusses occurs because "why" questions imply conditional acceptance. Recall from the section on suspending judgment that it is important to let speakers know they are accepted unconditionally, as full and unique persons who have the right to make their own choices according to their own values without having to defend themselves or measure up to some external standard. Unconditional acceptance is *not* an acceptance of all behavior or a statement that you agree with the person's choices or values.

The question "Why?" asks speakers to justify thoughts, feelings, and behavior rather than accepting that these exist and belong to individuals. Consequently, "why" questions put speakers on the defense and reduce the degree of trust, causing speakers to build defensive walls that block communication. Defensiveness happens regardless of the questioner's intentions to simply gather information.

Perhaps you can recall the following type of interaction with a parent.

PARENT: Why did you take that?
CHILD: I don't know.
PARENT: You don't know why you took it? How can it be that you did something without knowing why? Can you tell me why you don't know?
CHILD: I don't know.

Whether or not the child knew the reasons, she felt put on the spot, as though the parent's love or approval might be taken away if the truth were ever known. This is a situation of perceived conditional acceptance. The child fears punishment and rejection if she tells of greed and desire for the dollar, or a cookie, or whatever the item may have been. She believes it is better to look stupid than greedy and selfish.

◢ Exercise — Talk about It, Write about It

    A. Pair off with a classmate. Take turns talking about a time as a child when a parent, teacher, or other adult required you to justify your reasons for some action. Consider the following questions and talk about them.

       1. What were the "why" questions that were asked?

       2. How did you feel?

       3. Did you know the reasons but dared not tell them?

       4. Did you shut down to the degree that you no longer knew why you did what you did?

       5. If given the chance, now that the pressure is off, could you explain your actions?

       6. What feelings do you have as you refresh your memory about the event, the person, and the interactions?

       7. What gave this person the right or opportunity to press you?

    B. After you have finished the first exercise, talk about a recent experience in which you had to answer the same type of "why" questioning by a parent, boss, teacher, spouse, or other person. Talk about the similarities and differences between the situations and your feelings.

    C. Spend some time writing about the situation. This creates a permanent record, and you may develop further awareness of just how the intricacies of interpersonal power play out through the situation. ◢

People rarely think their reasons through carefully enough to be able to defend them later. Carefully phrased questions, with an attitude of acceptance, can help respondents think through their reasoning. Blame and punishment can come later; at the questioning point, we are looking for information and understanding.

Often, the information being sought through "why" questions can be better obtained by asking "what." Keeping the attitudes of empathy, postponing interpretation, and suspending judgment and asking, "what?" can elicit the desired information without provoking defensiveness. The respondent can give the details involved in decision making without being made to justify the choice, values, or soundness of the reasoning. Here are some examples of "why" questions alternately phrased using "what."

WHY	WHAT
"Why are you afraid?"	"What do you think will happen that scares you?"
"Why are you worried?"	"What is worrying you?"
	"What are you afraid will happen?"
"Why did you do that?"	"When you decided to do this, what were you expecting?"
	("What did you expect?" sounds like an obnoxious, "I told you so.")
"Why did you go there?"	"What did you expect to do or find that you chose to go there instead of staying home?"

Merely in terms of time, it seems easier to ask the "why" question. However, if saving time, energy, and frustration is your goal, I can almost guarantee that taking the time to rephrase the question will pay off. △

△ Exercise — Rephrase It

Take a moment to change each of the following "why" questions into a form that eliminates the power difference and increases the likelihood that you will get the information you are seeking.

1. "Why are you up so late when you have school tomorrow?"
2. "Why did you even bother to turn in a report with typographical errors?"
3. "Why didn't you get your grandmother to the airport on time?"
4. "Why do you wear such sloppy clothes?"
5. "Why didn't you call when you knew you would be late?"
6. "Why did you think I needed a self-help book?"
7. "Why don't you just go straight through college?"
8. "Why do you listen to the radio while studying when you know it creates distraction?"

9. "Why would you want to keep your watch set 5 minutes fast?"
10. "Why were you speeding again when you can't afford another ticket?"

QUESTIONS WITH OBVIOUS ANSWERS

Another type of question that incorporates a negative judgment is the question that has an *obvious answer.* Such a question is phrased so that only one answer is reasonable or acceptable and any other response makes the answerer appear foolish. These questions are often intended to force a respondent into the down side of a "one-upmanship" game.

Questions with obvious answers are very manipulative. They are related to "why" questions in that they also cause the respondent to feel defensive and unable to justify thoughts or actions. Some familiar examples from childhood are:

"Do you put your feet on the furniture at home?"
"If Jimmy jumped off a cliff, would you jump too?"
"How would you like that on the headline of tomorrow's newspaper?"
"Is this worth throwing your whole life away?"

Each of these questions brings in irrelevant situations as though they were comparable to the situation at hand. They are designed to elicit an obvious answer, forcing the respondent to think, for example, "If this marriage is throwing my whole life away, I *am* wrong and stupid, and I need to rely on your greater wisdom." The comparisons disintegrate because "The furniture in schools is steel and formica, not polished wood and padded cloth like at home", or "Jimmy didn't do anything nearly as drastic as jumping off a cliff, he stole a piece of candy." And on it goes with each question shown not to relate to what just happened. However, being on the defensive, the respondent may sense that there is something wrong with the comparison but not know how to challenge the authority figure who is "one-up" on the respondent.

Another type of question with an obvious answer is one in which the answer is stated as part of the question. Consider these examples.

"You do plan to get a police report on the accident, don't you?"
"People who really love their spouse will also accept the inlaws, won't they?"

"There's really only one right thing to do; resign your
position. So, what do you plan to do?"
"A kid with your ability would be a fool not to go to college,
right? So, which college are you looking at?"

Each question really creates or demonstrates a position of power for
the speaker by stating that the "only reasonable thing to do" is the
one presented in the question. The last example moves on to the
second question without necessarily waiting for an answer to the first.
Because the answer is obvious, actually waiting for the respondent to
provide it is unnecessary.

DOUBLE QUESTIONS

A fairly common problem with asking questions is the tendency to
pair them up. Recently, I came home from a night class to find my
2-year-old still up and running around. I asked, "What are you doing
still awake? Where's your Mom?" Obviously, the first question was
a waste of my breath to ask of a toddler. The second question was more
relevant and, once answered, might have allowed me to get the first
one answered.

Although part of the problem I had with my *double question* was
in asking a toddler, there are problems in general with asking double
questions. Whatever the motive or habit, double questions tend to confuse the respondent. Information gets lost because it is rare for both
questions to get answered. Be aware of your own tendency to ask
double questions, and separate them instead.

BOMBARDING

You may recall movies of World War II, with airplanes dropping hun-
dreds of bombs on a target area. *Bombarding* with questions is like being one of those planes tossing questions out like the multitude of
bombs. The problem is for the respondent, who has to deal with the
barrage of questions coming in. To the respondent, it feels more like
being interrogated than being interviewed or questioned.

People often resort, intentionally or not, to bombarding when they
perceive that their questions are getting them nowhere. Rather than
changing the approach, bombardment intensifies the current behavior
that has already been shown to be ineffective. Instead, choosing questions carefully and phrasing them to elicit the right kind of information will help you reduce the number of questions to be asked.

Compare these different approaches to asking whether someone is married.

> *Open:* What can you tell me about your intimate relationship?
> *Closed:* (Bombarding) Are you married? How long have you been married? How long did you date before that? Do you have children? How many? How old were you when you got married? How old were you when the children came? Did your parents approve of the relationship? Does your spouse get along with your parents? Have there been periods of instability in the relationship? (Worthington, 1992)

The closed questions could go on indefinitely, and the questioner may believe that the respondent is uncooperative; "Gee, getting information from you is like pulling teeth!" You can see that each of the closed questions, and several others, could have been answered with the single response to the first open question. Furthermore, there may have been information given in response to the open question that the questioner might not have anticipated when formulating a closed question.

Another way of reducing a bombardment is to acknowledge what is going on in the discussion. If you think the questions are ineffective, state so. If you think there is some other block within the relationship, it might be time to change topics for a moment to address these blockages.

Sometimes, people turn to bombardment as a display of power, either because there is a need to demonstrate real power or because creating a *sense* of power will make the interrogator feel less powerless or defensive. If you find yourself bombarding, check to see whether power is an issue. If you find yourself being bombarded with questions, then you might consider introducing power as a topic for discussion.

Exercise — Listen to Yourself

Keep track of your own questioning behavior for a few days. Keep a small notebook to record the patterns you identify and your impressions of them. Watch specifically for how and when you use the following:

    Closed questions
    Open questions
    Polite commands

Double questions
Bombarding

After gathering information for a few days, ask yourself these questions:

How did people respond to your questions?
Are you phrasing them well?
Do you question different types of people differently? How?
(For example, you might question children differently than adults.)
Are there patterns you were not aware of before?
What changes in the way you use questions will you make?

Identifying how you use questions will give you a greater ability to take control over when you question, how you go about it, and toward what ends you introduce questioning.   ⬧

In conclusion, we may all have reasons for doing things that seem sound to us but that we may not wish to defend to someone else. When presented with "why" questions, questions with obvious answers, double questions, or a bombardment of questions, most people are likely to lose trust in the questioner and attempt to withhold their thoughts and feelings.

If you find yourself asking "why" questions or engaging in any of the other problematic questioning methods, examine how you feel about the respondent and the situation. By using poor questioning techniques, you block effective communication and decrease the respondent's trust—exactly the opposite of your goal.

On the other hand, if you are asked questions in any mode described above, don't answer the question right away. Rather, point out your reaction to the questioning and ask what information the questioner is trying to elicit. You could say, "Are you asking me _____ ?" and suggest a rephrased question that is more appropriate.

⬧ *Exercise — Ask Some Questions*

Role play to practice using questions and probes. Find a partner, and decide who will take on the roles of speaker (tells personal narrative and responds to questions) and listener (uses active listening skills, uses appropriate probes and questions). As the speaker, select a scenario from Appendix B, taking a moment to think about the situation. Be prepared to expand on what is written in the scenario, trying to keep statements consistent.

The listener will begin as if the two people are meeting for the first time. As the listener, introduce yourself and ask the speaker's name (closed question). Then offer a broad open question allowing the speaker to state the nature of the situation. Use appropriate probes and questions to gather relevant information and to help the speaker discover answers independently, when possible.   ◬

Summary

Questions and probes are necessary ways of gathering information and clarifying what is understood. There are different ways of going about seeking information: subtle prompts use a minimal signal to encourage speakers to continue; clarification employs restatement and questioning to be sure information has been understood properly. Questions vary in their structure and the type of information they are designed to elicit. Closed questions gather specific details, such as "What city were you born in?" Open questions invite the speaker to talk at length: "What was it like to grow up in the city you were born in?"

There are several ways that using questions and probes can become problematic. Questions that ask "why" generally put the respondent on the defensive and may restrict the free and open flow of discussion. Questions with obvious answers create an environment of control rather than information exchange. Double questions and bombarding confuse respondents and may cause them to withdraw.

Generally, seeking information is a skill that can be powerful when used well and can create distance and disrupt communication and relationships when used poorly.

Skills of Reflection

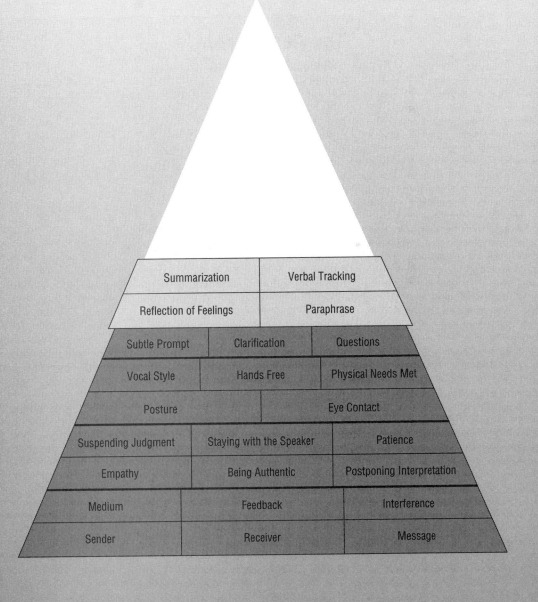

"The best mirror is
an old friend."

GEORGE HERBERT

Up to this point, the focus has been on creating a situation and style of listening that will help the speaker feel comfortable and talk. The next steps are to feed back to the speaker a condensed version of what has been said and keep the speaker on track.

The skills of reflection presented in this chapter are the reflection of feelings, paraphrasing, summarization, and verbal tracking. The active listener uses these skills together to assist speakers in gaining a better understanding of their own thoughts and feelings and to deepen exploration into verbal and emotional content.

The Skills of Reflection

Whereas the questions and probes of the previous chapter sought to gather more information and direct speakers to explore certain areas more fully, skills of reflection seek to provide a mirror that allows speakers to see their communication more accurately. Reflection of feelings and verbal content can help speakers focus more clearly on their own expressions.

People often report that psychologists and other skilled listeners seem to be psychic because they are able to ascertain clearly just what others are saying and feeling. This perception generally comes not from any psychic ability but from the fact that the skilled listener clearly reflects back the speaker's own words and feelings. People are not accustomed to being listened to and actually heard with intensity, clarity, or understanding. This is the power and importance of the skills of reflection.

REFLECTION OF FEELINGS

> **So when you are listening to somebody,**
> **completely, attentively, then you are**
> **listening not only to the words, but also**
> **to the feeling of what is being conveyed,**
> **to the whole of it, not part of it.**
>
> **J. KRISHNAMURTI**
> *You Are the World*

Reflection of feelings entails providing feedback to speakers about what they appear to be feeling. This skill has three basic purposes. First, reflecting feelings helps you gain clarity of the speaker's emotional state. Second, reflecting feelings can help speakers become more aware of their own emotional states; people often have difficulty attending to what emotions they are feeling at the moment and what impact the emotions have on them. Third, when speakers are caught up in a purely or primarily thinking mode, your focusing on the emotional content can help them get a more complete and accurate picture of their own experiences and reactions.

Many of us were raised with the expectation that part of being an adult is to become more rational and less emotional. To this end, we may have learned to reject or deny feelings and emotional reactions. This sentiment is expressed in such statements as "Don't be mad," "You shouldn't get so embarrassed about such little things," "There's no reason (emphasis on the rational side) to be jealous of your sister's accomplishment," and "If you let your feelings get the best of you, you'll lose the game."

Many people also learn to block desires. Certainly wants are driven by emotions; we want certain things or want to avoid other things because of the emotional connection we attach to them. Well-meaning people may block our desires. For example, as a child, I remember wanting to mix milk and water at the dinner table, and when I started to pour, my father said, "You don't want that." I remember being perplexed at the time, but I didn't have the mental ability to articulate the confusion, even to myself. Still, the experience stayed with me. The message behind my father's statement was, "I know what you want better than you do. Don't trust your own desires; check them with me first." Essentially, this statement was half true; I *did* want it, but I probably would not have liked it after taking a taste. It would have been more accurate for my father to say, "I don't think you'll like that."

Exercises — Feel It and Want It

1. Take a moment to write about 5 situations in which you remember being told what you should or should not feel. Perhaps there are situations in which you were doing the telling, or you might include the kinds of feelings that were taboo in your family. Address the following questions as you reflect.
   a. How did these experiences shape your ability to experience these feelings today?
   b. How might your learning to accept or reject certain feelings affect your ability to accurately identify them in others?

    c. How might your learning to accept or reject certain feelings affect your ability to have accurate empathy with others who are experiencing them?

2. Select some of the emotions you identified in Exercise 1 and make a plan to do something to heighten your ability to feel the emotions. Perhaps you will choose to watch a sad movie or read something that you strongly disagree with and get angry about it. It may help to enlist the help of a trusted friend, classmate, or teacher to develop a plan. The idea is not to force or pretend an emotional reaction but to increase the likelihood that you will begin to feel these feelings and become familiar with them.

3. Take a moment to write down five things you want that, at some time, you were told that you either should not or did not want. Write a brief statement about your reaction to each at the time and your reaction now.

4. Form a group of three classmates and discuss what you wrote in Exercise 3. Look for consistencies.

Reflection of feelings is often best presented as a guess or a hunch based on cues you observe in the speaker's statements and behavior. Some cues will be obvious, as when the speaker states feelings directly; other cues will be more ambiguous. Crying, for example, can be a sign of anxiety, grief, joy, sadness, or depression.

In Chapter 3, the purpose of the exercises "Label Your Emotions" and "Label Others' Emotions" was to familiarize you with your own emotional reactions and to learn the cues to use when guessing at others' emotions. In those exercises, you didn't get to ask whether your guess was correct. However, with reflection of feelings, you will be making educated guesses at people's emotions *and* getting feedback on your accuracy.

Keep in mind when reflecting feelings that the label a person gives to an emotion may cover a collection of emotions. Several subset emotions were listed in the exercise on labeling your emotions; for example, "frustration" may include fear, anticipation, aloneness, ambiguity, and restraint. Speakers may also use their own words that don't have well-known meanings. "I'm tweaked," for example, could mean a number of things. Be sure to verify that you understand just what the speaker means by emotional words. Sometimes, listing the subset emotions can help the speaker discover some new twist to the subject that was not considered before.

As you become familiar with a speaker, you will become more accurate in guessing and reflecting feelings. However, some people may get angry at your telling them what they are feeling. This is why it is important to pose a reflection as a hypothesis. Keep in mind the attitude of postponing interpretations of what you see. Following are some suggestions for how to phrase your reflection as a guess.

"You seem to be feeling . . ."
"I'm not 100% sure, but by your expression, it looks like you are feeling . . ."
"The way you are speaking makes me wonder if you are feeling . . ."
"You say you are feeling _____ . I wonder if there might be some _____ behind that?"

Each of these reflections gives your speaker a chance to deny the feeling if it is too risky to acknowledge or to correct you if you are wrong. Only experience can help you learn whether to pursue your guess, hold on to it for future reference, or accept that you were mistaken. The benefit of expressing your guess is that it may spark the speaker into exploring feelings more deeply. Either way, the result is the same—speakers get more clarity about their own feelings and you get more information about their feelings too.

When a speaker's emotional state is obvious, it is generally best to state the emotional content as being obvious. For example, if a woman comes into your office sweating, flushed in the face, agitated, and saying "I got to my car just as the parking meter ran out and they were there already writing the ticket! It wasn't expired a minute, but I got a parking ticket anyway! Aaaargh! Sometimes I hate this city!" You would appear ridiculous, inept, and insensitive if you said, "You seem upset." She is plainly upset; in fact, she's angry, mad, hostile. It would be more appropriate to say something like, "Gee, you're really mad about this."

In such situations, you may want to probe further to see whether this is an isolated frustration, connected with something else, or a sign that more vital issues are behind this degree of aggravation. You may follow the previous statement with, "So, is this all about the parking ticket, or is there something else going on, and this is just the last straw?" Of course, if you already know that the person has been under a lot of other stresses, you might introduce those issues and ask whether there's a connection.

Students of active listening often wonder how they will know what to say or ask or how to identify someone's feelings. This is where skill and art combine with practice and experience. The common advice is to "go with what the speaker gives you." That sounds fine to the experienced person, but it's not very reassuring for the novice. Let's take a look at how to discover embedded feelings that could be reflected. Read the following example and try to imagine a person saying it, paying attention to emotional content.

> SONYA: "I don't know what to do about my son, Miguel. I keep trying to tell him that he's okay. I try to let him know that I don't think of him as a failure, but every time I do, he seems to find something that he can turn inside out and interpret as rejection or disappointment. I don't think I show or feel anything negative. It's like the kid goes looking for something to confirm his own dislike of himself and then blames it on me."

You probably have some questions to help clarify what Sonya is saying and feeling, but stick to emotional content. Read the statement again, as needed, and write out what emotions you see and what words lead you to identify those feelings.

As I review the statement, I see a generalized concern for Miguel, along with fear of what might ultimately happen and a persistent sense of frustration throughout. Sonya seems to start off with hopelessness and confusion when she says, "I don't know what to do," and she becomes aggravated with either the situation or directly with Miguel when she mentions that "he can turn everything inside out." Sonya becomes uncertain and questions herself when she says, "I don't think I show or feel anything negative." Finally, she appears mildly angry at the end when she says, "It's like the kid goes looking for something . . .". These final words seem aggravated, though not particularly hostile.

Are there emotions that you identified that match what I described? Are there some that I didn't identify? Any mismatches? Mine are not necessarily right. Your interpretation may depend on how you imagine Sonya, and it may differ from mine because we have our own perspectives as different people. Without a doubt Sonya is frustrated, and I think this could be stated directly. However, some emotions are more subtle and require more interpretation on the listener's part. These might best be presented as tentative statements.

◬ Exercises — Identify and Reflect Feelings

Read the following scenarios, and use them for the exercises that follow.

SCENARIO 1: I think I deserved better. There were twelve students in that seminar course, and many I talked to said that I had often done the best work, contributed the most to the discussions, and had a grasp of the material that they could rely on during study groups. I don't think I deserved a B in the class when they got A's. Just because the instructor wanted us to be able to cite sources from memory, that doesn't justify grading down.

SCENARIO 2: My husband loves to cook. He takes care of the meal planning and preparation. I do the clean-up, but he's a fairly neat cook. I was relieved because I always saw the drudgery my Mom went through with meals. Growing up, I always just expected to have to deal with meals. But now I have time to spend reading to the kids, or doing something else.

SCENARIO 3: The book inspired me. I didn't realize how dead I was feeling until I started reading. It's like I could see myself in the author's words. She would describe how people become predictable like zombies, and there I was. She wrote about how love is love, until we get into a rut, then we forget to give and start to worry about what we are getting. And I looked at my relationship, I cried. I can't really say why, it was like a dawning, intense.

1. For each of the three scenarios, identify and write the feelings the speaker is expressing. Explain what in the statement shows you the feeling you identified. State which feelings you would feel certain enough to state as obvious, and which should be reflected as hypotheses.
2. Form small groups in class, and talk about what you wrote in Exercise 1. Discuss where you agreed and disagreed. Be sure to address what it was in each scenario that suggested the emotions you identified.

Emotions are an important part of most discussions. They can be used as a reference point when referring back to key moments in the discussion. They can also help speakers explore the full breadth of their experiences.

PARAPHRASE

A *paraphrase* is a brief restatement, by the listener, of the key points made by the speaker. It serves to provide feedback to the speaker and to gain clarity. Because speakers have the opportunity to hear what you understand from what was said, they can either agree or correct your understanding.

A paraphrase is appropriate whenever the speaker has made what appears to be an important point or when you want to make sure you understand what has just been said. To paraphrase, condense the key point into a sentence or two in your own words.

Once the listener has paraphrased what was said, the speaker has the opportunity to make corrections in what the listener understood. This way, both speaker and listener know that the message was received the way it was intended.

In the following conversation, Sam paraphrases what Cheryl has said.

CHERYL: A few years ago, I moved across the country—three thousand miles—because I was tired of where I grew up and needed to change the pace. I guess I just got itchy feet and took off.

SAM: So you moved across the country because you didn't like where you grew up.

At this point, Cheryl can agree or disagree. In the next example, she disagrees and Sam tries again to get it clear.

CHERYL: No, it's not that I didn't like home, I just wanted a different environment, some life in the fast lane.

SAM: So you just needed a change of pace.

CHERYL: Yeah, that's it.

Now Cheryl and Sam both know that Sam understands what Cheryl is trying to say. It is helpful periodically to take a larger chunk of the conversation and be sure that both people are clear about what has been said. This comes under the topic of summarization.

SUMMARIZATION

Summarization consists of putting together all the important bits of information gathered so far in the conversation. This is the point at which the listener may draw conclusions, make tentative interpretations, link what was said with other knowledge the listener has about the speaker, and receive feedback from the speaker to be certain there is a proper understanding. A summarization generally comes at the end of a discussion or at intervals during the discussion when either the topic is being changed or there is a need to bring several points together.

Following is an example using the conversation from above between Cheryl and Sam. The following summarization might come from Sam after Cheryl has talked for several minutes.

> SAM: Let's review this a moment. You moved across the country for a change of pace and were excited at the time. But recently you've been feeling sad, lonely, and out of place (reflection of feelings). You'd like to go home but you're worried that people will think you are a failure for giving up, even though they never even tried moving away from home. That makes you angry at them. So in all, you feel frustrated and stuck.

Sam's summary seems to have covered a lot, but Cheryl may have talked at length about all the details and unimportant points and spent some time trying to narrow her feelings.

Summarization is more than just a lengthy paraphrase. A good summarization provides clarity for the speaker as well as assures that the listener has understood correctly. In this sense, it helps the speaker better understand an issue, topic, problem, triumph, and so on. It may provide speakers with a new perspective on their own experience.

Paraphrase and summarization are related skills that tend to reflect content more than feeling, although both factors are generally included in poignant reflections. These skills are important in building relationships because they convey that someone is really listening. A natural outgrowth is trust, caring, and further disclosure. Deepening relationships then grow from the use of these skills.

Some caution must be taken in paraphrasing and summarizing. Both are powerful techniques, able to interfere as well as facilitate both the listening process and the relationship within which the listening occurs. Be conscious of the interview's purpose. In a conversation between friends, there is less need for summarization and more for maintaining a free flow of ideas, commentary, reactions, or other undirected interaction. In a business, therapeutic, or educational discussion, it is much more important to verbally check to see that what is said is being understood clearly. It is also more important in these latter situations for the discourse to maintain a particular direction.

The listener may focus on many parts of a discussion and paraphrasing may suggest that one idea is more important to focus on than others. By restating an idea, you reward the speaker for including it, attending to it, and perhaps even returning to it.

Learn and practice several ways of introducing paraphrases and summarizations. A cliché commonly used by novices is to continually state, "What I hear you saying is, . . .". The more you use this introduction, the more you will create distance between yourself and the speaker. Here are some alternatives.

"So, . . ."
"A moment ago you said _____ and now you added _____ ."
"I hear some inconsistency. Let me get this straight. . . ."
"I want to make sure I've got this right; . . ."
"Hold on. I think I'm lost. Do you mean . . . ?"
"Let's stop a second and piece this together. . . ."
"As I understand it, you're saying . . ."
"So, as you see it, . . ."
"It kind of makes you feel like . . ."
"I'm not sure I understand, but you seem to be saying . . ."
"So, from where you sit . . ."
"Your feeling right now is that . . ."
"Listening to you, it seems that you're feeling . . ."
"As I get it, you're saying . . ."

Try varying your introduction. At times, it might also be appropriate to launch right into the summary or paraphrase.

Some unskilled or novice listeners overuse these skills, sometimes because the listener doesn't know what else to say. When this occurs, I have generally found it best to disclose that I don't know quite what to say. The honesty of this statement and congruence with what is going on inside the listener lends credibility and gives the speaker a sense that the listener is human, authentic, and attentive.

Overuse of summarizing and paraphrasing can also result in "parroting." A parrot has nothing to say except what has been said in its presence. Continual mimicry of the speaker's words, feelings, and ideas can become annoying and create alienation. Consider the following ridiculous—but all too possible—example.

SPEAKER: When does the office close?
LISTENER: I hear you saying that you wonder when the office closes. I seem to be sensing some confusion there, or is it curiosity?

In this overstated example, the listener makes several mistakes. The first mistake was in saying, "I hear you saying . . .", which suggests the listener is a novice. Second, the speaker's question should simply be answered; content does not need to be rephrased or mirrored back. Third, the reflection of feelings is too tentative and irrelevant.

🔺 *Exercise* — *Practice It*

The only way to become skillful at paraphrasing and summarizing is through practice. In this exercise, pair off with a partner from class and discuss topics from Appendix A. Take turns being speaker and listener, paraphrasing and summarizing while being careful not to make the errors discussed above. Feel free to introduce your own topics, just be careful not to get so personally involved in a topic that you lose sight of the exercise.

After each turn, give constructive feedback to your partner about what was and was not included in the paraphrase or summary. Address how your partner's statements might have influenced what you said.   🔺

VERBAL TRACKING

Verbal tracking is a twofold process whereby a listener follows the speaker's progress through subjects and keeps the speaker on track. Speakers commonly stray from the main subject and go off on some tangent. This is fine in normal conversation, because the main intent may be to interact just to spend time together. A common result of straying off the subject is the familiar question, "How did we get onto this topic?"

In a helping situation, business meeting, or educational setting, there is a purpose to the discussion other than just for the sake of conversation. This purpose may be termed "topic of discussion." Sometimes, straying from the topic of discussion provides valuable information, but it also leads away from the main issues.

One job of the active listener is to help keep the speaker on the topic of discussion. First, remind speakers of the topic of discussion when they get sidetracked: "We were talking about how you can't seem to get along with your peers. Let's get back to that." Second, ask the speaker how the tangential subject relates to the topic. Perhaps it provides important information; perhaps it is just the next thing that came to mind. Digressions are especially likely to happen when people are emotionally upset, because they have difficulty organizing their thoughts and are likely to experience a barrage of ideas, each of which seems most important. By saying, "That sounds interesting (or odd, painful, and so on), and I wonder how this fits in with your problem about getting along with your peers?", the listener acknowledges the feelings and value of the tangential subject in its own context, then brings the speaker back to the topic of discussion by asking an open question about how it relates to the subject at hand. There may or may not be any relation, but the listener suspends judgment and gives the speaker a chance to pursue or discard the secondary topic.

Here is an example of a discussion in which one speaker goes off on a tangent.

TONY: No matter what I do, they seem to taunt me. They call me names and make fun of my work.

AN LI: Who are "they"? (Closed question.)

TONY: Roger and Julio.

AN LI: Okay. Specifically, what is it they do? (Open question.)

TONY: Well, they call me "stupid" and "twerp" and say my work isn't as good as theirs. (Tangent begins.) It's just like this one time when I was a kid and Tommy Dancy kept making strange noises at me and he was always pestering the other kids like me. I always felt like punching him, but I knew I'd get in trouble with the teacher, my folks, and everyone. (Tangent gets even further away from topic.) But there was this other kid who I got along with okay. He and I sort of were buddies and did things together. And one time . . .

AN LI: Hold on. (Listener expresses what is going on for her.) I'm getting lost here. (Summarizes.) You were talking about being a kid and not being liked by this Tommy kid, then you got onto the kid who did like you. (Expresses confusion over trying to follow the shift of topics.) I don't know where this is going. How does this relate to what we're here to discuss? (Open question.)

TONY: Yeah, I did kind of get distracted there. Well, I guess the part about that kid, Tommy, is that I feel the same way now. Like I'd like to punch him, I'd like to punch them. But of course I'd get in trouble and lose my job and all. So that's why I'm talking to you about it.

In this example, Tony certainly got sidetracked, remembering stories from childhood. Part of the stories related to the topic of discussion and part seemed to be lost meanderings. An Li was careful not to judge Tony's sidetracked story. She was thus able to learn that Tony's frustration had reached the point where he thought he could become violent and had turned to her instead. She now knows how important it is to Tony to get the problem resolved.

When interrupting to keep the discussion on track, be sensitive to timing. A well-timed verbal tracking, or any other interruption, comes when there is a natural pause between statements. It is perhaps best not to engage in verbal tracking too quickly; rather, make note of topic changes as the speaker talks. It is possible that the tangent is entirely relevant and the speaker will get back on track independently.

Some speakers don't provide any breaks, particularly when there is an intense emotional state; some of them seem to be able to talk on without ever taking a breath. At these times, you may need to interrupt, but be sure to do so with respect for the speaker. Interruptions can be initiated with "excuse me" or "pardon me" or with a more complete interruptive statement, such as, "Hold on. I think I'm getting lost here . . .".

However, even the most helpful interruptions become bothersome if they occur too frequently. Allowing speakers to discuss relevant tangent issues and come back to the main topic independently saves them from feeling directed instead of listened to.

Exercise — Keep It on Track

Pair off with a partner, and take turns talking about topics from Appendix A. The speaker should feel free to wander onto related topics, and the listener should make appropriate breaks into the speaker's commentary to keep the topic on track. Practice several times each. Be sure to provide constructive feedback to each other on the timing and the manner of the listener's attempts to keep the speaker on track.

Putting It All Together

It may be easier to learn each of the listening skills separately, but the art is in bringing them together to work for you as a unified whole—making you a competent listener. Putting it all together as a listener brings in each level of the pyramid presented so far and uses them all as background for listening. It means using each skill where appropriate to gain a deep understanding of messages as speakers intend them to be understood. It can also mean guiding speakers through their own confusion toward what they wish to express.

Summary

In this chapter, the skills of reflection were presented, showing how they rest on the preceding levels of the pyramid. These skills are powerful tools for deepening the level of disclosure and for strengthening relationships. Be sure to use them carefully so you don't end up parroting the speaker and interrupting unnecessarily.

Following is a list of each skill of reflection, with a brief statement of its purposes and an example of its implementation.

1. *Reflection of Feelings:* to help speakers notice and attend to emotional states; to be clear that the emotions you see are those the speaker is feeling.

   EXAMPLE: "From what you've said, you seem to be sad and confused about this."

2. *Paraphrase:* to reiterate key points in your own words, provide feedback, and gain clarity.

   EXAMPLE: "So you moved for a change of pace."

3. *Summarization:* to review what has been said, provide feedback, and gain clarity; to bring together major points, emotional content, and a culmination of related issues that were discussed.

   EXAMPLE: "Let's review this a moment. You moved for a change of pace, and now you've been feeling sad, lonely, and out of place. You'd like to go home, but you're worried that people will think you are a failure for giving up. That makes you angry at them. So in all, you feel frustrated and stuck."

4. *Verbal Tracking:* to follow the speaker through topics; to keep the speaker from straying off the subject.

   EXAMPLES: "This problem at work reminds you of other times when you were a child."

   "How does what you've just said relate to what we were talking about?"

Speaking to be Heard

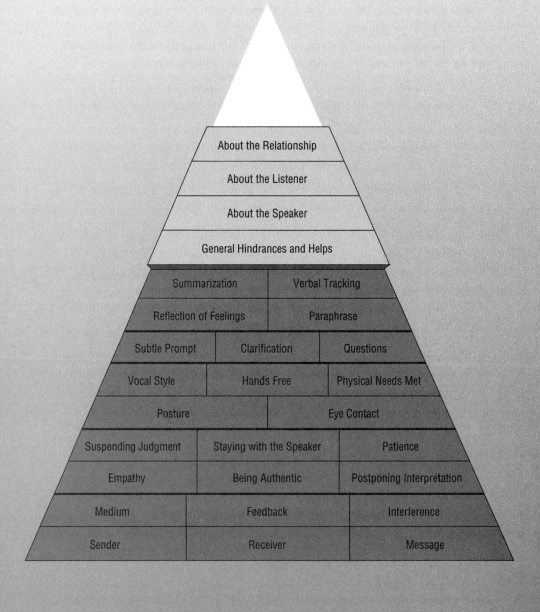

**"The time has come,"
the Walrus said,
"To talk of many things . . ."**

LEWIS CARROLL

*Through the Looking Glass
and What Alice Found There*

In this chapter, you will learn skills to help you speak in a manner that increases the potential that your intended message will be received accurately. Although there is no guarantee that the message you intend will be the one your listener perceives, you can take active steps to increase the probability of getting your message understood. This chapter addresses the issues and teaches the skills that constitute those active steps.

General Hindrances and Helps

Now that we have examined how to listen to a speaker, let's reverse roles and examine what it means to be a speaker who is trying to be heard. Being heard is more than merely having the message sensed by the ears or even perceived by the brain. It needs to be assimilated by the mind, taken into the consciousness of the listener. So far, you have been learning how to effectively assimilate a message. However, most people we interact with will not have had training in active listening. Regardless of how competent your listener may be, effective communication requires that you speak in a manner that is most likely to get your intended message received.

HINDRANCES

Before delving into the right things to do to be heard, it may be beneficial to examine common errors that amateur helpers make when communicating. Worthington (1992) provides a convenient list of unhelpful responses. Look over this list while reflecting on whether you would be helped or feel listened to by someone who responded in such a manner.

Ordering, directing, commanding.
Warning, threatening.
Moralizing.
Advising, giving solutions, suggestions (even if they
    are requested).
Lecturing, teaching, giving logical arguments.
Interpreting, analyzing, diagnosing.
Reassuring, sympathizing, consoling.
Withdrawing, distancing, humoring.
Attempts to distract. (pg. 57)

Read through the list again, thinking of a time you approached someone or someone approached you and the response fit the description of the above terms.

I won't belabor the point by giving a commentary on each item or line. I have done most of the things on the list and found them to help me avoid much real contact with the other person, and I am pretty certain that most readers will have had a similar experience. Now, let's turn to the helpful characteristics of speakers who are heard and how they construct their messages.

HELPS—BEING DIRECT, SPECIFIC, AND CONCRETE

**He was wont to speak plain
and to the purpose.**

WILLIAM SHAKESPEARE
Much Ado About Nothing

Directness refers to the ability to come straight to the point. People who speak directly tend to let others know exactly what they have in mind, with a minimum of beating around the bush. *Being specific* means that the speaker is explicit and definite about what is being said. Finally, *concrete* statements operate in the realm of tangible, physical, observable, and measurable terms. Behaviors and actions are concrete, whereas thoughts are not.

A direct statement states specifically and concretely what one is thinking, feeling, or experiencing. It is not hidden by forced and false etiquette or any other decorations that might cloud the message. Feelings are most clearly expressed by being direct, stating the feeling in words instead of only through the more subtle cues of intonation and body language. Compare the following two statements made by a young boy who wants to accompany his older brother and friends to a ball game.

"Gee, I like ball games too."

Or,

"Can I come along?"

The first statement is indirect and only hints at what the younger brother desires. The second statement comes directly to the point. Here are some more examples of direct statements:

"I feel sad and lonely when you tell me I can't go with you."
"I was really glad that you did the dishes."

> "When you smoke in my office I get angry and pay more attention to the smoke than to what you are saying. So please don't smoke in here."

Notice the parts of each of these statements that are consistent with each other. First, in each statement, the speaker takes responsibility for feelings and makes this plain by saying "I." Second, there is a mention of specifically what the other person is doing that prompts the speaker's reaction. Third, where necessary, a request for a specific behavior change is made, as in the last example that asks for concrete and measurable changes.

The specific aspects of these statements include the speaker's emotions and the listener's behavior that engenders the speaker's emotion. Returning to the first example, let's examine what is being said.

> "I feel sad and lonely when you tell me I can't go with you."

The speaker states his emotion specifically and in the context of the situation—"I feel sad and lonely"—rather than using the word *lousy*, which is very general and nondescript. The listener's behavior that triggers this reaction is "you tell me I can't go with you." These items are specific and concrete; the older brother knows just what he is doing that leads to this reaction and what to do to prevent it if he chooses.

The speakers of each direct sentence above are aware of exactly what they are feeling and wanting from their listeners. They are able to take responsibility for their feelings and needs and communicate them. Their behavior is authentic. Furthermore, being in touch with what they are feeling lets the speakers bring up the matter before it is only one of a series of perceived offenses. This way, they have prevented a blow-up of anger that results from an accumulation of little offending incidents.

These speakers are being authentic because their direct statements reveal what is truly going on inside them. This requires a level of trust in both the listener and in oneself. There is always the risk that the listener will be rejecting or say "tough luck" to some request or statement. When this happens, at least it is because of the listener's unwillingness to address the issue, not the speaker's.

Only one of the examples given makes a request of the listener— the one in which the listener is asked to stop smoking in the office. The others state the speaker's reaction but make no request. Thus, listeners can make a change in behavior if they do not wish to continue eliciting that reaction in the speaker, but no demand for change was made. When this fact is understood between the speaker and listener, open, nondefensive communication is more likely to occur.

This mood can be set by beginning the statement with, "I am not asking you to change this; I just want you to know how I feel about it. You can change if *you* want to."

Say what you are thinking and feeling. Your feedback and requests will be more effective if they are concrete. Before speaking, consider what it is you want the receiver to hear. Precisely *what* things are you concerned with that you want to communicate to this person?

△ Exercise — Making Concrete Statements

In this exercise, be sure to make your comments direct, specific, and concrete. If there is a suggested request for future behavior, make sure it conforms as well. Refer back to the text as necessary.

Rewrite the following statements and questions so they are more likely to be heard. Use your imagination to create the scene in which each statement is made. You should be able to rewrite each statement two or three different ways.

"When are we having dinner?"
"The carpet would look good if it were vacuumed."
"People who get their work done on time get promotions."
"I like yogurt too."
"I haven't got time to deal with this right now."
"I've enjoyed spending time with you."
"We're having a little social get together next Friday."
"I can't concentrate with this noise!"
"How can someone with your schedule do all this?"
"We only have three tickets, and your name's not on any." △

About the Speaker

**When you want to recognize and understand
what takes place in the minds of others,
you have first to look into yourself.**

THEODOR REIK
Listening with the Third Ear

The speaker, in this chapter, is you. You will bring your own personality, personal history, values, goals, desires, and anxieties to the situation in which you are speaking to be heard. Some things that will help a listener remain open to your comments are making "I" statements, engaging in appropriate self-disclosure, and avoiding the temptation to give advice.

MAKING "I" STATEMENTS

In the United States, most English speakers are taught to speak in the third person because saying "I" sounds self-centered and is too personal. *Making "I" statements* means breaking this cultural taboo and saying "I" when talking about yourself.

Making "I" statements is perhaps the single greatest method of reducing defensiveness in listeners. By shifting the focus onto their own thoughts, feelings, perceptions, and reactions, speakers avoid the problems of blaming and labeling. "I disagree" is a very different message from "you're wrong." This is not to say that either side is always equally right; when the topic addresses a matter of opinion or evaluation, then discussing agreement is more accurate than rightness.

When listeners hear the word *you*, they tend to protect themselves. Most of us have experienced being labeled, blamed, and told who we are or how we are feeling. People often perceive such statements as attacks and feel the need to defend themselves. The operative word here is *perceive*. Whether it is there or not, if the listener perceives an attack, interpersonal walls will be erected and communication becomes ineffective. Despite the intrigue people have for fortunetellers and handwriting analysts, in communication, people do not like to be told how they are feeling, what they are thinking, or what they do or don't do.

There are two basic types of "you" statements. The first makes some claim or comment directly about the listener. Let's turn to some examples.

"You're the kind of person who . . ."
"You never really did like . . ."
"You're not taking this stuff very seriously."
"You don't appreciate me."

In these examples, the speaker is making the grave error of telling another person what that person thinks, feels, or values. Although the speaker may be right about some of these, the rightness of the statement will not penetrate the defenses erected to protect against how it feels to be labeled or spoken for.

The second type of "you" statement displays a speaker's defense. This presents information about the self within the context of a generic "you"; that is, the word *you* does not necessarily refer to any individual listener but to everyone in general.

> "You know how when you have a bad day you just can't get out of bed."
>
> "Sometimes you really like to get someone mad."
>
> "You just can't get through this without caffeine."
>
> "People don't know how to take a compliment."
>
> "You have to cheat in business, it's that simple."
>
> "When your kids have a tantrum and embarrass you in public, you just want to scream and hit them."

Clearly, these statements tell more about the speakers than they tell about the "you's"being referred to. In these examples, the speakers have maintained the cultural rule against talking about oneself, but by doing so, other people have been labeled or spoken for. The level of accuracy in these statements about other people is low because the statements are about the speaker; they are just given a socially appropriate disguise.

The Rule Against "I". "Don't talk about yourself." "Don't start every paragraph of your essay with *I,* even if it is about yourself." These are just two of the messages that have indoctrinated many people in the United States against talking about themselves. Generally, people learn to avoid saying *I* but continue talking about themselves by substituting the word *you.*

The problem with using *you* to say *I* is that the former retains its original meaning as well. Here is an example of how saying "you" disrupts communication. The statement, "You just can't get through this without caffeine," was actually made to me once during final exams. I didn't know about the person who said it, but I got through everything without caffeine because it kept me awake all night. And I resented the assumption that the work I accomplished was due to the drug instead of my own efforts.

As this was very likely a statement about the speaker, it would have been best to make the statement "I just can't get through this without caffeine." In this statement, the speaker is far more direct, honest, and more likely to communicate successfully. People who are incorrectly included in the generic "you" would not end up building walls of resentment, and the listener would learn something about the speaker.

The adjustment to speaking with "I" statements may be difficult initially because it is hard to break cultural taboos. However, the payoff will be more accurate, personal communication between speaker and listener. Messages will be heard and understood the way they are intended and listeners will not be as likely to shut down.

What's Going on with Me? To learn to make "I" statements—to speak personally—it is necessary to know your own thoughts and feelings. Being aware of internal processes and expressing them reflects the attitude of being authentic. Asking yourself *"What's going on with me?"* helps you understand your own investment in what you are saying.

Using *I* generally seems threatening because it reveals part of one's own personality, values, or beliefs. The speaker is no longer lost in the crowd but singled out, vulnerable to personal scrutiny and rejection. I am unable to hide behind a mask of "the generalized everyone" when I state that it is myself who does or thinks whatever is being disclosed. Returning to the caffeine example, by saying *you* the speaker hid his own need for caffeine behind the generalization that "everyone does it." This way, he did not have to feel embarrassed about his own use of the stimulant. Had he said, "I need caffeine to get through this," he would have risked letting others know he relied on caffeine.

Taking Responsibility. By making "I" statements, speakers also *take responsibility* for, or "own," what they say. Consider the difference between these two descriptions.

> SPEAKER 1: "You feel so good when you drive fast around the corners. You feel the wheels grip the road, then start to slip. Ah, the excitement runs so high, you're on cloud nine."
>
> SPEAKER 2: "I feel so good when I drive fast around the corners. I like to feel the wheels grip the road, then start to slip. Ah, the excitement runs so high, I'm on cloud nine."

In the first statement, the speaker assumes that the listener also enjoys the feeling of being on the verge of losing control of the car. In the second statement, the speaker takes responsibility for feelings and owns the reaction without trying to project it onto someone else or assuming that the listener also enjoys the thrill ride. Whether the particular thrill is shared or not, the listener is not put in the position of being cast in the same light as the thrill-seeking speaker.

When you know what is going on within yourself and take responsibility for your statements by saying *I* when you mean *I*, what you say is more likely to be heard because the listener will not need to be defensive; defensiveness is a type of interference. Listeners may not like or agree with what you are saying, but they are more able to hear your message because you are only talking about your own feelings, thoughts, reactions, or experiences.

Taking responsibility for what you think, feel, and do is an active process. Taking responsibility is also reflected in word choice; active words give the power of control to the speaker, and passive words give this power to outside influences. Consider these two statements:

"I can't sing and dance on Main Street."
"I won't sing and dance on Main Street."

The first statement, although quite a common way of phrasing things, is not only passive, it is inaccurate. The person claims some disability by stating *can't*, reflecting inability rather than choice. In the second statement the speaker takes responsibility for making an active choice not to participate. She says she *will not*, which reflects the exercise of her will rather than some force over which she has no control. Other active words include *will*, *choose*, *decide*, and *feel*. Although the outcome of each statement—and each way of thinking that lies behind it—may be the same, clearly the second speaker is more in control of her life and behavior.

Of course, we are supposed to be excused from responsibility if we are unable to do something. This way, saying "I can't" should let us off the hook, whereas people are free to apply social pressure if we say, "I don't want to" or "I choose not to." It may take some personal fortitude to keep your needs and priorities in mind and communicate them clearly without the other person either taking offense or trying to break them down.

SELF-DISCLOSURE

Self-disclosure can be defined as the intentional verbal revelation of thoughts, feelings, and experiences. This does not deny that what we say, our phrasing, how we dress, how we move, and in fact everything we do tells something about us. However, what we are concerned with here is voluntarily and intentionally telling others about ourselves when our primary role is to listen. Hepworth and Larsen (1993) relate two types of self-disclosure. First is our responses to others; "I agree with your opinion," or "I don't understand what you just said." The second type of self-disclosure is telling anecdotes about ourselves or disclosing personal struggles, issues, or information. These anecdotes might be prefaced with, "I had a similar experience with my kids . . ." or "When I was in college, my roommate and I got into a major disagreement when . . .". Self-disclosure on the part of an active listener has many potential benefits. It can encourage the speaker's self-disclosure, partly by modeling and partly because of the understood cultural rule of interpersonal reciprocation: that we are expected to match the types of behavior others have shown us. Self-disclosure can also help build the other person's trust in your experience and compassion.

On the down side, self-disclosure can sometimes reduce trust and rapport. Disclosing your own successes and strengths may make you seem "perfect" or "superior" to the speaker. When disclosing personal struggles, the other person may wonder how you could help someone else when you can't even fix your own life (Hepworth & Larsen, 1993). People can also get sidetracked by their own stories. Be cautious about telling every related anecdote that comes to mind.

Be sure that self-disclosure in a helping relationship is relevant to the situation, relevant to the person, and well timed. Consider whether your reactions or experiences will contribute to meeting the other person's needs. Consider whether the person is ready to hear and integrate your comments and whether you are at the right point in relationship development for the disclosure. Well-timed disclosures build relationships; poorly timed disclosures create alienation.

Consider the example of a client who enters an appointment complaining that her daughter just moved out of the house 2 days after her 18th birthday. The client complains of wondering why it happened, feeling as if she failed as a parent, and having fears that her daughter won't be able to make it on her income. Your observation that the client has talked for 6 months about wanting her house and her life back after her daughter moves out might not be well timed until after the client has had time to work through the initial shock. You might also want to wait before disclosing that you moved out at 19, got into debt, and moved back in with your parents for 4 years while paying off debts.

Requests for disclosure are another issue. When a client, student, or employee asks about personal information, there is usually some underlying motive other than simple curiosity. In my first job out of college, I was a tall, thin young man working with older women in a weight-loss clinic. I was often asked whether I had been overweight. The simple answer was "no." However, what was behind the question was whether I could relate to the plight of these women. When I disclosed that I was underweight, had trouble gaining weight, and had suffered social ridicule because of this reverse weight problem, the women generally trusted that I could understand their problem. This facilitated rapport and allowed us to get on with the task at hand.

Sensitivity to cultural differences is also necessary in using self-disclosure, because members of different cultural backgrounds expect different levels of disclosure. In some cultures, a person who does not disclose is seen as maintaining a professional stance, whereas in others this is perceived as being secretive and untrustworthy. It is best to learn about the cultural norms of populations you will be interacting with (Gudykunst & Kim, 1992).

In general, it is best to use self-disclosure sparingly, when well timed, and with concern for the relevance to the person, situation, and type of relationship. It is equally important to define and maintain your own boundaries (as discussed in Chapter 3).

GIVING ADVICE

"Take my advice—I'm not using it." Many statements like this have been made about the odd phenomenon of *advice giving.* In a listening situation, and particularly in a helping relationship, giving advice is generally more hurtful than helpful. Successful advising is a real skill and an art.

The first thing to do is assess exactly what is being sought. Many who seek advice are actually looking for approval; others are looking for someone to offer solutions to their problems and, consequently, to take responsibility for their personal decisions. Occasionally, someone is looking for input about how to handle a situation from others who are more experienced in that area. A wise advice seeker comes looking for options and alternatives to consider that would not have been identified independently. Wise advice seekers are rare. Let's look at some of the problems that might arise when giving advice.

Perhaps the greatest ethical consideration in advice giving is that clients in a counseling relationship will become dependent on your recommendations. In an attempt to make the "right" decision or to live in a manner that meets with approval, people will learn to always seek advice. It is often easier for all concerned either to simply tell advice seekers what to do or to do it for them. However, offering solutions or fixing other peoples' lives is contrary to the goal of helping people learn to make decisions, live, and work independently. Furthermore, if someone follows your advice and things turn out badly, they can blame you and evade all personal responsibility. Similarly, if you give advice but don't follow it yourself, you are a hypocrite.

Even when people seek out advice, they may be paralyzed with feelings of helplessness, hopelessness, or inability to make real change. This resistance to any form of even the best-intentioned help is most apparent—and most frustrating—when the person draws an advisor into the "Yes, but . . ." trap. No matter what solutions are offered, an acknowledgment is followed by a myriad of reasons that it can't possibly work. Inevitably, the one seeking advice comes away claiming that the advisor was of no help, again responsibility is shifted to the advisor. There is no way out of this trap except to stop and turn the request for advice back on the seeker.

Being turned to for advice can be very flattering. I know that I feel a sense of pride and importance when students or friends come to me for advice. However, this also provides a great opportunity for me to make a fool of myself. Be wary! Instead, turn the request back onto the seeker, asking questions such as:

"What would you recommend if someone else came to you
    with this problem?"
"If you had to decide right now, which way are you leaning?"
"What draws you to each option?"
"What frightens you about each option?"
"What do you stand to gain or lose?" or, "What's at risk?"
"Have you really already made up your mind and are just
    looking for confirmation?"

Another technique I have found useful for helping others recognize their preference is to flip a coin. Heads is yes, tails is no, or whatever answers are relevant. If the coin lands and you haven't decided, then you do what the coin says. It is surprising how quickly people can decide when they start hoping for heads.

Turning the advice back onto the other person helps you avoid being blamed later, helps to prevent the "Yes, but" trap, and keeps you from appearing hypocritical. It helps people learn to help themselves.

Exercise — Avoiding Advising

Role play advice seeking with a partner, taking the roles of speaker and listener. The speaker will seek advice, the listener will attempt to deal with the request in light of the information presented in this section. The listener can select from the following situations, or make up his or her own.

"I've been thinking about signing up with the Navy . . ."
"I want to go out with this woman/man who is married . . ."
"I can't decide on a college major; what do you think I'm
suited for?"
"I found out my boss did something unethical . . ."
"My best friend wants me to get him (her) hired where I work,
but she (he) doesn't take work seriously . . ."
"I have $1,000 saved up. Should I take a vacation, invest,
save . . .?"

About the Listener

Speaking to be heard requires being sensitive to the listener. Advertising agents are the true artists at packaging a message for a specific population. It is fairly easy to discern the target population for different products by looking at the ads. For example, auto companies direct their advertising for large, expensive luxury cars toward the upper middle class—or anyone who wants to be identified as belonging to this class. The ads for small cars are directed toward young people with modest incomes; advertisers call these people "economy minded," which carries positive connotations.

Also, different products use different ads depending on where the ad will be placed. The same product may be advertised in both *Up the Corporate Ladder* and *Back to Nature*; however, it is almost certain that the ads in each magazine will be designed for appeal to the readers' very different value systems suggested by the titles of these magazines.

Advertisers are experts at tailoring a message for a specific audience because they analyze the target population carefully. Because there are millions of dollars to be made or lost through advertising, those who create ads need to research the demographics of areas, learn the values of those who will receive each type of media, and take surveys of preferences. Ad agencies learn about the audience and create an ad to appeal to this audience. Nothing is overlooked; the colors, the medium (TV, radio, newspaper, magazines, billboards), the models, and all other aspects are intentionally brought together to maximize the product's appeal to the target population. This intensive attention to communication increases sales.

So much for advertising; the point is that effective communicators pay attention to who their audience is and how to speak to that audience. If everyone were to develop some sensitivity to the special needs of others with whom they communicate, we could avoid many problems that end up being blamed on misunderstanding and poor communication.

SKILLS IN COMMUNICATING

Your listener's level of skill in communicating may be very different from your own. It will be very discouraging to approach others with a statement about how you feel with no intention of asking them to change, only to have them think you are being manipulative. Furthermore, they may have no conception of what it means to make "I" statements or any of the other techniques presented in this book. In most situations, it is generally best to maintain your own directness and realize that the other person is not always going to hold the same attitudes that you do; different people are just that—different.

THE LANGUAGE OF THE LISTENER

S. I. Hayakawa (1962) frequently wrote about the confusion involved with different vocabularies and different definitions of the same word. Consider the difficulty associated with the terminology of wines. The connoisseur refers to a wine as "dry," which appears to the uninitiated to be a contradiction in terms. Once the novice learns that "dry wine" means "not sweet," the two can communicate more accurately.

One problem with technical jargon and slang is that the definitions are not familiar to the rest of the population. In the English language, there is great redundancy in words. Even when there seems to be a "perfect word" for something, it is rare that there are no other words to describe the same thing. Although it may make sense to use the "best" word, this often leads to a breakdown in communication.

For example, once I tried to work on my car using a repair manual. By the time I got halfway through the second paragraph on tune-ups, I was lost. There were pictures showing what to look for, but I could not match the pictures to the corresponding areas of the engine. Being unfamiliar with the terminology of automotive repair, I found the manual to be worthless for my needs. We could not communicate because of the interference created by an unfamiliar vocabulary.

A general rule of thumb is to make sure your definition matches that of the other person. Be quick to doubt your assumption that definitions are the same. When there is even a hint that your definitions might not match, ask, "What do you mean by the word _____ ?"

This will clear up a misunderstanding before it can have too dramatic an effect.

It is generally best to match the language used by others if you understand the terms accurately. Imagine visiting the doctor complaining of a bump on the back of your head. On the next visit, you would probably be perplexed if the doctor asked, "How is the contusion in the occipital region?"

Loaded Words. A *loaded word* is one that carries strong connotations. Generally, although not always, the same strong connotations are shared by most of the population of a culture or region. The emotional component of the connotation is triggered when the word is used, creating a change in the listener's focus of attention. The emotions become the focus instead of what is being described or what follows the loaded word. Some examples of loaded words are *abortion, the Holocaust, prejudiced,* and almost every word that is used as an insult.

People generally construct euphemisms for loaded words. For example, a century ago, the word *idiot* was a diagnostic classification for people with low IQs. Later, the term became too loaded as an insult, and the term *retarded* was introduced. Ultimately, this word, too, became loaded and the term *developmentally disabled* was introduced. Recently, *disabled* has come under fire as an undesirable and inaccurate label, needing to be updated again.

Often the most frank and direct words are very heavily loaded, but it may be necessary to use some of these terms to clarify the nature of a problem, especially during a counseling session. For example, a victim of sexual assault may say something like, "He took my clothes off and then he grabbed me." The problem here is to determine how and where he grabbed her. Cultural taboos, coupled with the stress of the situation and the formality of speaking to a professional person who is a stranger, may make the victim reluctant to use "vulgar" or graphic language to describe the assault, even when this is the only way to communicate the details to the counselor.

An obvious problem with using loaded words in a helping or counseling situation is that they may create distractions from the actual matter at hand. Although it may be necessary initially to risk the use of loaded words to discover the facts, it is important to be sure that trust has been established before introducing too many of them. Once the facts and definitions are established, communication can be achieved using the familiar euphemisms, which may reduce the anxiety attendant to the discussion and thus help keep it open.

SEEKING CLARIFICATION FOR THE LISTENER

Earlier, we discussed clarification as a means of making sure that the listener had a clear understanding of what the speaker meant and felt. As an active speaker, it is also important that your listener have an accurate understanding of *your* thoughts and feelings. The following statements suggest ways of seeking feedback to be sure the listener has a clear message.

> "Okay, now can you tell me what I just told you so I know you got it?"
> "I want to make sure you understand me correctly, so tell me what you understood."

It's as easy as that. The hard part is remembering to ask and then recognizing when there is a possible misinterpretation of the message. It takes some practice to know when a message may be coming across with less than perfect clarity. Merely asking, "Did you understand that?" is ineffective, because the respondent will only be acknowledging that the message was understood as the respondent understands it, not necessarily as the questioner intended it.

One way to learn foresight is to study your own faulty communications through hindsight. The last time you gave driving directions over the phone, how much trouble did the person have in finding the destination? Ask where the trouble happened and how your directions were misleading. What part of the directions were unclear? How could you have given the directions to make them clearer?

When giving directions, either with friends, family, or at work, pay attention to *how* you present the directions. Try not to change anything else at first, just pay attention to how you present the information. Then see whether the listener had trouble understanding your message.

◢ *Exercise — Give Directions to Draw It*

With a partner or by dividing the class into pairs, designate one person as the sender and the other as the receiver.

With the partners sitting back to back, the sender draws a few overlapping geometrical shapes on one piece of paper without letting the receiver see the drawing. The receiver then tries to duplicate the picture by following the sender's verbal directions, such as "Draw a vertical line about 5 inches long, then . . .". The receiver is not allowed to see the original, and the sender is not allowed to see what the receiver is drawing until it is finished. If this game doesn't give you an idea of how errors in messages and directions can occur nothing will. It can also be quite comical!

Several modified versions of this exercise demonstrate how different levels of feedback affect the accuracy of communication.

1. Let the receiver ask questions.
2. Let the sender watch the receiver draw and give corrective instructions.
3. Allow both partners to speak freely.

The purpose of feedback is to promote accurate understanding, and we have already sufficiently discussed many aspects of feedback. We now turn to a common problem in achieving accurate understanding.

The obstacle to accurate understanding is the phrase, "Any questions?" Because listeners understand their own perceptions of your message, what they perceive might not be what you intended. Take the following example.

DAD: Don't use the car Saturday night. Any questions?
SON: Nope.

That interchange seems plain enough—or is it? Does it mean that the son can use the car Friday night? He asked no questions about Saturday night, but he may assume that he can use the car at other times. Perhaps the father's intent was that his son could not use the car that Saturday night in particular and could not use it any other time as well.

The way to increase the likelihood that a listener has understood your message the way you intended it is to use cumulative feedback. In this process, the speaker asks the listener to embellish what was heard by adding cumulative details in each successive statement. Using the car borrowing example above, let's look at how this might work out.

DAD: Don't use the car Saturday night. Let me be sure we understand this the same way. Can you tell me what that means? (Requesting cumulative feedback.)
SON: Sure, I don't get to use the car this Saturday.

DAD: And what does that mean about Friday? (Requesting cumulative feedback.)

SON: Well, I don't usually get to use it without permission anyhow. So I guess that doesn't change.

Or,

SON: I'll just use it Friday night instead.

With either final response, the father knows how the son interpreted the message and can prevent a misunderstanding. It took a bit of questioning on Dad's part, but in the long run it was probably well worth it to avert potential misconceptions. The questioning can seem like "pulling teeth" until both persons are familiar with the concept of cumulative feedback. It requires skill and patience to have this work out simply and easily.

THE IMPACT OF CULTURE

Culture is the collection of social rules, attitudes, behaviors, skills, and technology of a group of people. There are differences between people of different cultures, and these cultural differences affect communication. If your listeners are not of your culture, they are abiding by other rules and norms that you are unlikely to know (Gudykunst & Kim, 1992). For example, most people in the United States have a "rule" against saying *I*. Others have rules against conflict of any sort. Others have rules against direct eye contact between people who are considered unequal.

Care must be taken not to assume that all members of a given group share the same cultural experience. Even among those who were born and raised in the same neighborhood, people of different racial, ethnic, religious, or other backgrounds often have very different experiences and outlooks on life and relationships. For example, Prater (1992) points out that "Black children live in two worlds—the African-American community and a predominantly white society" (p. 94). Changing behavior to meet the norms of each group is an adaptation similar to, although perhaps on a grander scale than, the changes most people make between behavior at work and behavior among friends.

People have different assumptions about underlying verbal and nonverbal communication. Differing assumptions of appropriate role behavior can have disastrous results in a work setting. Consider the example of a U.S. supervisor who asks a Greek employee how long it will take to complete a job. Although the U.S. supervisor thinks of this question as asking for employee participation in decision making, the Greek employee may think, "His behavior makes no sense. He is the boss. Why doesn't he tell me?" (Gudykunst & Kim, 1992, p. 140).

Unfortunately, it is beyond the scope of this chapter to look closely at the broad and important topic of culture. However, there are organizations on many college campuses that address issues relevant to groups in the minority. Local chambers of commerce may be able to refer you to similar organizations in the community.

About the Relationship

Getting to know the person you are communicating with requires attention to potential influences on your relationship. Factors that affect the relationship between speaker and listener is the topic of the next section.

RELATIONSHIP DIMENSIONS

The relationship between a speaker and a listener can take many forms. The participants could be friends, acquaintances, lovers, spouses, boss and employee, or a combination of types. We form relationships for a variety of reasons and with a variety of different people.

Some of the dimensions that affect interpersonal relationships are context, time, intimacy, and control (Adler, Rosenfeld, & Towne, 1992). *Context* refers to the setting within which the relationship forms, such as the family, workplace, school, and car pool. Relationships are affected not only by the amount of *time* spent together but also by how that time is spent and how much of that time together is by choice. Simply being around one another may create a relationship, but this does not mean the relationship is valued. *Intimacy* involves a depth of self-disclosure. If we talk about a lot of topics but only on a superficial level, there is not much intimacy. Finally, *control* describes a spectrum of dominance and submission. It is rare that both partners in the relationship are equal in this dimension. One partner usually takes and is given the lead in decision making, relationship definition, and other activities. Furthermore, it is not always the one who appears to be in control who actually wields the power.

Every relationship involves some conflict and a control issue. Some relationships are created with stated differences in control, such as a parent and child, but eventually at least one partner will want to change the balance of power and control. Ways of dealing with conflict and the emotions that build up over time are usually less well stated, but eventually all conflict comes to the surface.

Emotional Baggage and Unfinished Business.   These concepts from Gestalt psychotherapy describe the backlog of feelings a person may be carrying around. The idea is that people save up or hold on to emotional memories from the past (Corey, 1991). All this baggage can get in the way of clear communication. If emotional baggage gets aired indirectly during a discussion, the listener is likely to feel unfairly attacked.

For example, Tim has felt for a long time that Nancy has lost interest in him and has been looking at other men. He then may be unduly critical when confronting her for forgetting to buy milk. Because of the emotional baggage Tim carries, his trust is lowered and he harbors extra hurts and angers. Even though Nancy's forgetting milk may rate a mention or even a confrontational discussion, it does not rate the type of blow-up that results when Tim's other hurts and angers come out indirectly during the discussion of the milk.

Fear, Trust, and Imbalance of Power.   These three issues are discussed together because they tend to occur together. It is rare for a person to be afraid while feeling in control. It is when this control is weak or threatened that fear arises and trust becomes an issue.

Let's look again at the example with Tim and Nancy and their disintegrating relationship. Tim feels lack of trust in Nancy because of his fear that she is interested in other men, but he is afraid to confront this issue. This puts Nancy in a position of power over Tim because she is in control of her affection and attention. So, Tim fears Nancy, fears the situation, trusts neither Nancy nor the situation, and thus feels powerless with regard to Nancy's impending decision to leave or stay. All three factors operate together.

Other relationships with an imbalance of power include supervisor/employee, teacher/student, counselor/client, and parent/child. Because of the imbalance of power inherent in these relationships, the person with less power may be afraid to confront the other person about anything at all. In effect, the lack of willingness to confront demonstrates a lack of trust. Of course, some people cannot be directly and successfully confronted. The subordinate person may be reading the situation correctly and is therefore wise not to confront if keeping the relationship intact is more important than addressing a particular issue.

On the other hand, sometimes the person with more power is the one who needs to do the confronting. Nancy may decide to talk to Tim about what is happening, a supervisor may talk to an employee about continual tardiness. In these situations, it is best for the powerful person to keep in mind that the listener is likely to be afraid of losing something and will have high anxiety. The person confronting should begin by acknowledging the listener's possible anxiety and either validating or dispelling the sense of alarm: "Howard, I know you're probably anxious about this meeting. You have good reason to be."; or, "Howard, I know you're probably anxious about this meeting. Well, you can relax, it's not so bad."

By disclosing your knowledge of the person's anxiety and giving some clue about the severity of what is to come, you demonstrate that you care. This demonstration builds trust even if it does not make the ultimate message any easier to say or hear.

CARING CONFRONTATION

It is rarely comfortable to confront another person. We are likely to wonder how we will perform under the stress and how the other person will respond. The fear is that we will ignite some interpersonal explosion. The ideal outcome is that the person agrees and says, "I've been trying to bring this up myself." Although there can never be a guaranteed outcome, caring confrontation will help to minimize the problems.

Caring confrontation involves the assertive statement of one's observations, reactions, emotions, needs, and desires within the context of the attitudes of active listening. There is a big difference between caring confrontation and being "nice" or sugarcoating an angry or hurtful statement. Caring confrontation is a direct, honest, and open way to say uncomfortable things.

There are two general types of situations for confrontation. The more familiar one involves telling someone that there is a problem. Whether the problem rests in the relationship, a person's behavior, a rumor, or anything that disturbs a person, the discussion is likely to be heated and uncomfortable for all participants. The other type of confrontation involves providing feedback to a person about inconsistencies between statements, speech and behavior, verbal and nonverbal messages, or other actions. This type of confrontation often occurs in helping situations as when a counselor points out that the client smiles while saying, "I am so angry." Regardless of the type of confrontation, the same principles apply toward creating a successful interaction.

Being specific, concrete, and authentic are the keys to sending a confrontational message that is likely to be heard while creating the least amount of defensiveness in the listener. You should carefully consider specifically what you need to say to the listener, what your motivation is, and what emotions are involved. If you are feeling intensely angry, think about how this anger might cloud the directness of your statement. Is there a backlog of anger-producing situations that need to be discussed, or is there truly just the one item? Remember that being direct in your confrontations will help prevent an emotional outburst on your part or a defensive reaction from your listener.

When engaging in caring confrontation, think about the attitudes of active communication. Each of the attitudes for active communication come into play in confrontations.

1. Do you have *empathy* for your listener? Are you taking into consideration how it will feel for the other person to hear what you are about to say—from that person's point of view?
2. Are you *being authentic* by confronting the person on the matter? Do you have some hidden agenda you might not be fully aware of? A good way to determine your authenticity is to notice whether you are irritated with the person in general or very fearful of the confrontation.
3. Are you willing and able to *postpone your interpretations* of what this person has done? The situation may not actually be as it appears at first sight. Be willing to hear the reasoning behind the behavior that caused your reaction.
4. Are you *suspending judgment?* Your confronting the situation means that you had a reaction that probably includes a judgment of your listener's behavior. The key to successful confrontation is to be able to suspend the finality of your judgment while learning the motives and reasoning behind the other person's behavior.
5. It is very important to *stay with the speaker* during a confrontation (that is, when the other person is speaking). To postpone interpretations and suspend judgment, it is necessary to listen closely without trying to formulate your next statement. When this quality is lost, the people in a discussion do not hear each other and the discussion disintegrates into an argument.
6. *Patience* with both yourself and the other person, along with perseverance, helps bring a mutually satisfying conclusion to the discussion. The willingness to "hang in there"

even when emotions start to boil will help clear the air instead of clouding it even further. Perseverance when confronting does not mean preventing any breaks from the action. Rather, it is often wise to take some time, even if it's a day or so, to regroup and calm down. However, perseverance does not mean that you let the discussion fade away just because it is uncomfortable to return to. Unless there is a return to the discussion, both people will likely feel an uneasiness about themselves, the situation, and the other person; it is like waiting for the other shoe to drop.

A good way to make sure the discussion gets finished is to make a contract to finish it. One person could say, "I'm getting flustered here and need a little time to go over what we've said. Let me take a break and we'll continue this at 2:00." This sets a specific time at which both people know the discussion will be continued. Of course, it is important that both people agree to the contract.

By keeping these attitudes of active communication in mind, the confrontation will more likely come from caring and will be perceived by the listener as such.

Another aspect to examine in a caring confrontation is the end goal. What do you truly want to get from the discussion? If everything goes your way, how will things be different after you are finished? Finally, are these expectations realistic in light of the other person's needs and perspectives?

It may take quite a bit of soul-searching to find the answers to these questions, and your listener will probably pick up any hidden agendas, conscious or not. Remember that it takes trust to provide a climate of openness sufficient to allow a mutually caring confrontation. If there is a serious challenge to trust, it will take more work to arrive at a satisfactory conclusion to the discussion.

Determine whether your goal is to have the other person change behavior or attitudes. Any time you tell someone to change, he or she is likely to feel threatened and defensive. Depending on *how* they are confronted, people may be open to hearing your messages or trying a new way of behaving. The art is in creating the presentation.

If your goal is merely to let the other person know how you are feeling, your job is easier. It helps to tell people directly that you do not expect or need them to change, then say what you need to say. Remember that with caring confrontation, the emphasis is on the *caring;* if you care about the other person and about the relationship, you are more likely to succeed and have all parties end up feeling good about themselves, the other people, and the resolution.

Summary

Speaking to be heard effectively improves communication, and well-constructed speech helps listeners understand the intended message with greater clarity. As a consequence, listeners will be more willing and able to risk open discussion because the effort the speaker makes toward understanding demonstrates concern, and demonstrated concern promotes trust. Trust in turn promotes greater openness and risk, and the process builds toward open, honest, and direct communication.

Some ways to communicate your messages more effectively are to be direct, concrete, and specific while avoiding advising, offering solutions, lecturing and judging, and diagnosing or interpreting. Make "I" statements when talking about yourself and your thoughts and feelings. Use self-disclosure sparingly and judiciously in a helping situation, and always make sure it contributes to the interaction rather than detracting from it. Stay attuned to what listeners want and need. Giving advice and offering solutions takes responsibility away from listeners and may alienate them.

What the listener brings to the interaction will also affect how you construct and send messages. Take into account the listener's skills in communicating. Assess their language proficiency so you talk neither above nor below them. Make sure they understand your messages by using clarifying skills beyond simply asking, "Do you follow me?" Always be mindful of the listener's culture and how this might affect the accuracy and efficiency of communication.

The relationship dynamics will also affect your ability to achieve your goals. Power differences and confrontation require special attention so the context they create does not interfere with message transmission too dramatically.

In the next, and final, chapter, we will examine some of the jobs and situations in which active communication skills can be applied. Ultimately, at the peak of the pyramid, you will model and teach these skills and attitudes to others.

Applications

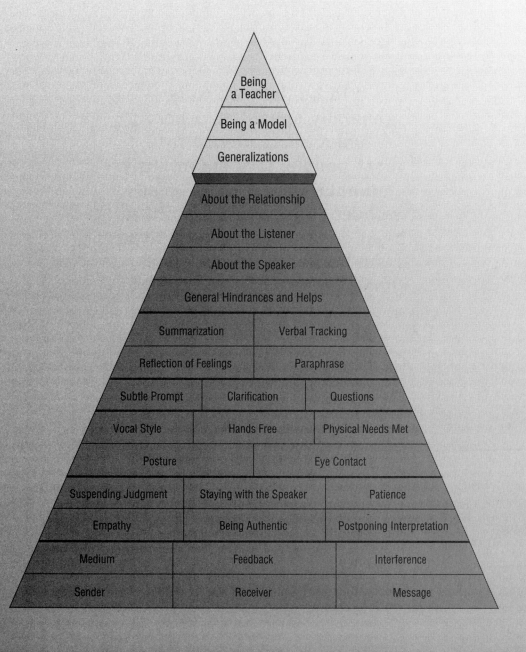

"To be listened to is, generally speaking, a nearly unique experience for most people. It is enormously stimulating. . . . It is small wonder that people who have been demanding all their lives to be heard so often fall speechless when confronted with one who gravely agrees to lend an ear."

ROBERT C. MURPHY

Psychotherapy Based on Human Longing

In this chapter, you will be shown the value of applying the principles of active communication to all the varied areas of your life. Although you should not limit yourself to this list, the areas presented are family and friends, therapy and counseling, crisis counseling, business, sales, education, and allied health professions. You will also be given a few ideas about teaching active communication to others through being a model and a teacher. Although this is our shortest discussion, the driving force behind it should be considered the most important; skills are of little value if they are never applied.

Generalizations to Life

In psychology, the term *generalize* can be defined as "going from the particular to the general; to extend" (Rathus, 1993, p. 3). The intent of this chapter is to encourage you to extend what you have learned about active communication and generalize the skills and attitudes to every situation you encounter. Any time there is an exchange of information or ideas, there is an opportunity to use active communication.

Another aspect of generalizing is spreading the knowledge to others. Later in the chapter, we will address how you can teach these skills and attitudes and be a model for others.

To help the attitudes and skills take a firm foothold in your habits of communicating, you will need to use them continually. Even when you are sitting with friends at lunch, you can practice. It is hard to be a part-time active communicator and still be a good one. You probably won't be able to hold the attitudes in only some situations and still have them be authentic.

However, this does not mean that you always have to take the role of "the active communicator." You don't need calling cards saying, "Have Ears and Mouth—Will Travel." Sometimes you may want to dominate a conversation; sometimes you'll want to be passive. This is all a part of leading a whole and balanced life.

If you work to integrate the concepts of active communication into your life, you will undergo a change in the way you view things. This is not to suggest a complete "personality change"—you won't go from Dr. Jekyll to Mr. Hyde—but you will experience a change in some parts of your lifestyle. People may begin talking with you because you appear to care sincerely, because you listen and understand. Take advantage of this opportunity to sharpen your skills. On the other hand, pay attention to your own inner sense of whether you want to be involved with this person in this role. Protect yourself from becoming everyone's crying towel and getting burned out as a result. To take care of others, you must also take care of yourself.

Before discussing teaching the skills and attitudes, we will first explore some situations to which active communication skills apply.

FAMILY AND FRIENDS

**One friend, one person who is truly
understanding, who takes the trouble to
listen to us as we consider our problem, can
change our whole outlook on the world.**

ELTON MAYO

The family is a prime place to use active communication. This is the place where people first learn communication patterns. If there is to be a "natural" movement toward healthy communication patterns, it will come from people learning these attitudes and skills from infancy through teaching and modeling.

Taking the time and energy to listen actively and see beyond the superficial aspects of communication in a family will help smooth out conflicts because people will feel truly heard. Also, the listener will be able to break patterns that keep family members from achieving the trust and intimacy they so often desire yet consider unattainable. If nothing more, active communication can start the development of healthy family patterns.

Among friends, active communication helps strengthen relationships that can become very important to maintaining all participants' self-esteem and support systems. Active communication will help people understand the struggles and concerns of each other's lives from the other's viewpoint. This makes friends feel close and intimate.

THERAPY AND COUNSELING

Many, if not most, readers are probably "therapists-in-training." However, whether therapist or client, it is useful to employ active listening skills. A large part of the therapist's job is to listen. By using the skills and by holding the attitudes presented here, therapists assist clients in evaluating and seeing their lives and choices in a new light.

Teaching clients effective communication skills can be an important and therapeutic intervention in itself. Helping clients set aside their own assumptions while they hear what another person is actually saying can help them smooth out many conflicts in relationships and other interpersonal interactions. Teaching clients to speak with "I" statements shows them how to take responsibility for their own thoughts, beliefs, needs, and wants and helps them communicate their own thoughts more accurately.

Depending on the therapist's theoretical orientation, several other therapeutic techniques may also be employed, such as interpretation, contracts, and homework assignments. However, one thing common to all good therapists is their ability to listen with understanding (Rogers, 1961).

CRISIS COUNSELING

Crisis counseling is almost definable by the fact that it utilizes a minimum amount of time to make a maximum amount of impact on the speaker. Good communication skills are essential in this highly volatile situation. It is the job of the crisis counselor to:

1. Understand the plight of the speaker as the speaker experiences it.
2. Assess this person's needs.
3. Take action to help this person get these needs met quickly.

Accuracy in steps 2 and 3 depends on the counselor's ability to perform the first step. Gaining this understanding requires a solid ability to listen and empathically understand the client's words and feelings.

BUSINESS

Many situations arise in business settings for people to stop, step back from the immediate situation, and listen to each other. This action helps keep bosses and employees happier and more productive. When actively communicating, a boss is more likely to hear an employee's remark before it becomes a formal complaint. By paying attention to an employee's statements, the boss will also be more attuned to the subtle clues that indicate when an employee is either unable to perform some aspect of the job or has out-performed expectations and deserves recognition.

An actively communicating employee, on the other hand, will be better able to understand a boss's directions by summarizing them and confirming understanding through cumulative feedback. The employee will also be able to communicate better with co-workers and may end up in a position of leadership.

SALES

A simplified explanation of selling involves two steps. The first is making a product or service available; the second is convincing people to choose one product over another or even to choose to purchase a product at all.

Convincing people to purchase requires assessing their needs or wants and presenting a product or service that best fills these needs. Active listening assists by helping the salesperson get an idea of what the customer needs. Active speaking helps the salesperson make a presentation of the best match between product and need.

EDUCATION

Teachers must be effective communicators to succeed at their jobs. As an instructor, my greatest struggle is not with the course content but with constructing messages that will penetrate students' preconceived notions and affect the way they understand the world around them. Educators must be sensitive to the verbal and nonverbal cues that students send. Teachers must provide fast and beneficial feedback so students know how they are progressing. A good educator is, by definition, a good communicator.

ALLIED HEALTH PROFESSIONS

Doctors, nurses, physical therapists, dentists, and the whole range of health professionals need to use an active approach to communication in order to perform their jobs. When a patient comes for consultation, an astute health professional will have empathy to see the person behind the complaint and to construct questions so as to elicit the most complete and accurate responses possible. Given the hurried schedule of health professionals, it is all the more important to be precise in speaking, listening, and meeting the psychological needs of patients. Patients who believe they are being heard will be more satisfied with the service and more willing to follow through and seek aid when necessary.

Incidentally, it is as important for the patient to use active communication as it is for the health professional. Patients particularly need to apply the techniques of asking questions, making probes, and speaking to be heard. Patients must be sure to seek clarification on any directions or medical advice.

Thus, active communication is an approach to life. There is no circumstance in which a person cannot apply these attitudes and skills.

Everywhere there are people whose communication styles could bear great improvement. Watch what is going on, and see how you could improve someone's understanding of what has been said.

Teaching Active Communication

An important part of applying active communication to your life is to pass it on to others. An easy way to teach the skills is to give them a book such as this. A more effective way is to model the skills; just use them and let people see what you are doing. And the most effective way is to teach by explaining the skills and attitudes while modeling active communication.

BEING A MODEL

A wonderful way to bring people's attention to communication style is by modeling. Using the skills and holding the attitudes of active communication will demonstrate to others that there is a better way to communicate. They will begin to notice a pattern to your style and will comment on it. Then you can be more direct about what you are doing because they will be motivated to listen. From there you can tell them more about just what it is you are doing and how to do it.

BEING A TEACHER

I consider teaching active communication to be the peak of the pyramid. Teachers have the greatest possible influence on the greatest number of people. However, whether you are in front of a class or in a casual setting with friends, everyone influences the people they come into contact with; in this manner, everyone is a teacher.

If someone asks about your active communication skills, tell them what you are doing. Describe and demonstrate the skills, discussing attitudes, interference, listening skills, speaking skills, and how they have worked in your applications. One way to teach is to break the attitudes and skills down individually and show how they build on each other. You may want to use the pyramid analogy presented in this book. Give examples and role play with others. However, keep in mind that there are some inappropriate situations for teaching. For example, an argument is not the time to say, "See, you could do this too." This is likely to be perceived as condescending and patronizing.

In conclusion, application is the pinnacle of achievement for any skill. However, you have to do the groundwork before reaching this peak. Apply what you learn as you progress, but remember that regularly refreshing the skills and reevaluating the attitudes will help you continue to grow toward greater effectiveness in communication—and this can only make life easier.

Summary

The point of this chapter was to emphasize that active communication skills should be generalized to every part of life. Active communication can be applied in any situation and at any time. It is also important to pass along the "secrets" of active communication. Being a model is probably the easiest and most effective way to teach others the value of active communication.

Consider this parting comment from *Bring Me a Unicorn* by Anne Morrow Lindbergh.

> **Then I want to sit and listen and have**
> **someone talk, tell me things—their life**
> **histories—books they have read, things they**
> **have done—new worlds! Not to say anything—**
> **to listen and listen and be taught.**

Topics for Exercises and Discussion

Following is a collection of topic suggestions to use with the exercises and discussion sessions. Feel free to add your own. These are designed only to give you ideas if you can't think of anything to talk about.

A dream of flying
A favorite daydream
Infatuation, your first puppy love
Separating love, infatuation, and lust
What the moment of death might be like
Yourself at 80 years old
Yourself in 20 years
The technological item you most dislike
What your clothing says about you
Your favorite time of day
What was so special about the best car you ever had
A time you moved far away
An ideal vacation
Unisex rest rooms
What it means to you to be a man
What it means to you to be a woman
Which character you would be if you were in *The Wizard of Oz*
What do you do when caught in traffic
The ideal college course
The ideal job
If you suddenly earned twice your current income
If you suddenly earned half your current income
A scary dream you had as an adult
A scary dream you had as a child
What made the difference between your best boss and worst boss

A good book you've read lately
A good movie you've seen lately
Cartoons you've enjoyed
How handy you are (or aren't) around the house
Cooking—how, when, and what you like to cook
Something funny you saw
Something you wish you had recorded on videotape
A playground fight you got into as a child
A moment of self-discovery
A personal "high"
An artistic talent you have
An emergency in which you had to stay in control
Being in the hospital
Public speaking
If you were a cartoon character, which one you would be
Your goals for the next year
Your goals for the next 5 years
Research for a term paper that taught you
    something interesting
Being on time; being late
How organized or disorganized you are
The benefits of simple living
What "enough money to live on" means
The first day of kindergarten
Something with great personal but no monetary value
A disaster—where you were and your reaction
    when you heard
Something you did on a dare as a child
A favorite childhood hero, and what made the hero special
What you admire in people
What makes you angry about people
The most important life lesson you learned
What competition does for (or to) people
Things that make you unhappy
A major turning point in life
The silliest thing you are afraid of that can't really harm you
Something you would like to stop doing
How you tend to think, feel, or act in a crisis
A time you were robbed
Some things you are really good at

Role Play Scenarios

Following is a collection of scenarios for role play exercises. Use these as a starting point and build from them. Different groups can use the same scenario without duplicating the role play because each participant brings a unique personal experience and interpretation to the role play. Don't feel limited by these topics; add to them, introduce your own, and role play scenarios you have experienced that you want further insight into. The scenarios are separated by general fields, although they often certainly overlap.

These scenarios are designed to be gender neutral and are worded from the perspective of the speaker—the person seeking advice, input, or other help. Most are created so that the speakers can initiate the discussion without the listeners knowing any more than what job roles they hold (supervisor, teacher, and so on).

Remember that the listener's job is to elicit further information and encourage the speaker to express thoughts, emotions, concerns, and reactions, not to provide answers or give advice.

Human Services, Social Work, and Counseling

1. You were just raped. A friend told you to call the police, but you don't want to be asked all their questions, and you don't want this to be part of the public record. You feel that "This isn't supposed to happen to people like me!"
2. Last month you engaged in a high-risk sexual activity, and now you feel sure you contracted the HIV virus. You can't sleep or eat and can barely perform your job. An HIV test won't tell you anything conclusive until at least 6 months have passed.

3. You took your teenage daughter to the emergency room for appendicitis, and it turned out to be labor. When you asked your daughter who else is responsible for the baby, she said, "Nobody. She's all mine, I'll never tell who the father is."

4. After reading about all the recent earthquakes, you feel sure that a major quake will strike your area. You think about dying or being trapped under concrete for days. You've started driving long distances out of your way to avoid going over or under bridges. You're looking for a one-story apartment.

5. You feel anxious all the time. You don't know why; you just feel a constant dread that "something terrible" is going to happen.

6. Everything is a joke. You have trouble taking anything seriously. Your spouse is getting annoyed that you don't react to anything except to joke about it. Your last counselor suggested that it is a defense mechanism, so you changed counselors.

7. You feel stuck between two job offers. One is a dream job but requires you to relocate. The other isn't quite as perfect but pays a little more and lets you stay where you are.

8. You have two sons, and your spouse always wanted a daughter. You are tired of the "infant–diaper–late-night" routine and want to move on with your life. This issue has become the first real conflict in your relationship with your spouse.

9. Your spouse just served you with divorce papers and moved in with a new lover. You were taken completely by surprise and fear that you are inadequate, unworthy of love, and will never realize your dream of having a spouse, children, a house with a picket fence, and all that comes with having a family.

10. Last night, your live-in lover hit you again, bruising your back. It happened once before but your lover swore it would never happen again. You don't want to lose your lover but feel you can't continue in this type of relationship.

11. You are an elderly resident of a nursing home. You want to tell the patient advocate that you think someone has been taking things from your room. However, you are

very worried that things aren't really stolen but that your memory is failing and you are misplacing things. You don't want to falsely accuse anyone, but you also don't want to be robbed while you sleep. You've heard about this type of abuse and are afraid that if you tell, you will be treated badly by staff of the nursing home.

12. You live in a retirement home. You believe that your children are handling your finances improperly and are spending your money on themselves without your authorization. You gave them power of attorney so they could pay your bills directly, but now you think things are not being handled properly.

13. In a disaster situation (earthquake, tornado, or hurricane) your 8-year-old child has been trapped in the debris of your collapsed apartment building for 3 hours. The rescue crews report they can hear that people are alive in the debris, but they are out of reach. Your own child's condition is unknown.

14. During a flood, you have come to the emergency relief shelter set up by the Red Cross. Your family is all together and healthy, but your house is completely flooded with mud and filthy water. Little, if anything, is salvageable.

15. You are a teenager in a group home for emotionally troubled adolescents. You are angry about the lack of privacy in the bedroom you have to share with others.

16. You are a developmentally disabled adult living with your parents and attending a sheltered workshop. You complain to your case worker during a home visit that some new kids in the neighborhood are chasing you with their bicycles and calling you "retard," and you are afraid and humiliated.

Allied Health Professions

1. You are a 14-year-old girl whose doctor just told you that you're pregnant. You are afraid to tell your parents.

2. You are a 72-year-old terminally ill patient who wants to be released from the hospital to die at home. Your family is reluctant to let you do so.

3. You are a 33-year-old diagnosed with cancer 3 days ago. You want to refuse treatment because you are afraid that you will just die a longer, slower, more painful death than if you let nature take its course. You have no religious or moral restrictions against treatment, only personal reasons.
4. You feel anxious much of the time. Your job and family cause you stress, and you want medication.
5. Since the death of your spouse 6 months ago, you have felt lost and depressed. You don't want medication.
6. You are an elderly person approaching your doctor because you want to try LSD and marijuana before you die. You figure you have nothing to lose because you're going to die soon enough anyway, and you won't pass genetic problems on to your kids. You have always been "good" by avoiding illegal drugs and now want doctor's blessing and advice.
7. You have been married 12 years. You want an HIV test because you think your spouse has been having an affair and you fear you have been exposed to HIV.
8. You are an elderly person with failing health, angry at your daughter and her husband for putting you in a nursing home. You are upset that they seem to think you are losing your mind and are unsafe around the children.
9. You are spending your first night in the hospital, awaiting abdominal surgery. Suddenly, you feel very sure that if you are anesthetized, you will never wake up.

Education

1. You have been called into your college instructor's office for being late to class three times in a row. Your assignments have also been turned in late. You are taking 15 credit hours and working 30 hours as well. You think you have too much to do.
2. During a parent/teacher conference, your child's teacher expresses concern that your child isn't doing schoolwork. You are a single parent with a low income, and cannot afford school supplies.
3. You are a high school student, telling your teacher, "My parents are too strict . . .".

4. You are a high school student who has cheated your way through school this far and never really learned to read. You wonder, "What's wrong with me? How can I fix this before it's too late?"

5. You say to yourself, "I always get A's and at least B's. I failed this test because I don't see where school is relevant, so I quit bothering. I just don't care any more."

6. You are a junior high school student. Your peers are pressuring you to have sex.

7. You are a junior high school student who tells your teacher, "A fellow student pulled a pocketknife out during class, held it to my leg, and said I had better do his (or her) homework too. I was also told that if I told anyone, I would be attacked and cut after school."

8. Some friends of yours cheated during the last test. They wanted you to cheat too, but you said you didn't want to. You aren't sure who they all are, but if they're caught, they will be sure you are the informer.

9. You honestly did your paper, but over the weekend, your house caught fire and most of the house burned. You have some remnants of the report in a plastic sandwich bag, clearly burned and apparently soaked from fire hoses.

10. You are a gay high school junior or senior. Other students are not aware of your sexual orientation, but seem hostile to you because they make jokes and wisecracks. You would like to go to the prom, but you are worried about being derided for bringing a same-sex date. You have considered attending with an opposite-sex date, but don't want to mislead anyone and don't really want to spend this important evening with someone to whom you feel no attraction.

11. You are terrified of graduating (to move on to either junior high, high school, college, or the work world). You don't know if you'll be able to make friends, succeed or survive the new challenges. You fear that you are entering "the big time now" and are just not ready.

12. You are a girl in high school who is upset because your parents disapprove of your clothing. Your father calls them "slut clothes" and your mother says you look like you're asking to be raped, or at least treated like a sex object. You think they're stylish and you've spent a lot of time and money on those clothes.

Business

1. You are 43 and want to start your own business. You have ideas that are not listened to where you work, and you're not appreciated and are passed over for promotions.
2. You are called into your supervisor's office after being 30 minutes late to work three consecutive times. You just can't get your kids off to school any earlier; they resist.
3. You pull all the weight in your work crew. Everyone else works only enough to get by, and not always that, so you end up doing all your work *and* some of theirs. They have come to expect that you will finish whatever part of their quota they don't get done, because bonuses and reprimands are given to the crew, not individually. You are angry but afraid to confront the crew because you fear their retaliation.
4. You took home about $200 worth of office supplies—a stapler, pens, paper, and such—and you want to confess.
5. Your company has some illegal policies, and your job requires that you engage in some illegal actions. You explain to your supervisor that you trust (him or) her and don't want the company hurt by federal investigations. You also don't want to lose your job or go to prison.
6. You want a transfer to a different department because your co-workers discovered something about your past. It is none of their business and doesn't affect your work, but they tease you about it. You have confronted them and asked them to stop, but they taunt you all the more for supposedly not being able to take a joke.
7. Your company just became "computerized." You don't like computers, don't trust them, and refuse to use e-mail. Because you won't read your e-mail, you have missed two important meetings. Your boss is angry, but you plan to hold to your anticomputer position.

References

ADLER, R. B., ROSENFELD, L. B., & TOWNE, N. (1992). *Interplay: The process of interpersonal communication* (5th ed.). Fort Worth, TX: Harcourt Brace Jovanovich.

AXTELL, R. E. (Ed.). (1993). *Do's and taboo's around the world* (3rd. ed.). New York: Wiley.

COREY, G. (1991). *Theory and practice of counseling and psychotherapy* (4th. ed.). Pacific Grove, CA: Brooks/Cole.

COREY, G. (1995). *Theory and practice of group counseling* (4th. ed.). Pacific Grove, CA: Brooks/Cole.

COREY, G., COREY, M. S., & CALLANAN, P. (1993). *Issues and ethics in the helping professions* (2nd. ed.). Pacific Grove, CA: Brooks/Cole.

FAST, J. (1970). *Body language.* New York: Pocket Books.

FROMAN, R. (1986). How to say what you mean. *ETC.: A Review of General Semantics, 43,* 393–402.

GUDYKUNST, W. B., & KIM, Y. Y. (1992). *Communicating with strangers: An approach to intercultural communication.* New York: McGraw-Hill.

HALL, E. T. (1973). *The silent language.* Garden City, NY: Doubleday.

HAYAKAWA, S. I. (1962). How to attend a conference. In S. I. Hayakawa (Ed.), *The use and misuse of language.* Greenwich, CT: Fawcett Publications, Inc.

HEPWORTH, D. H., & LARSEN, J. A. (1993). *Direct social work practice: Theory and skills* (4th. ed.). Pacific Grove, CA: Brooks/Cole.

MAY, R. (1953). *Man's search for himself.* New York: Norton.

MCLUHAN, M., & FIORE, Q. (1967). *The medium is the massage.* New York: Bantam Books.

MORRIS, C. G. (1993). *Understanding psychology* (2nd. ed.). Englewood Cliffs, NJ: Prentice-Hall.

MORRIS, D. (1977). *Manwatching: A field guide to human behavior.* New York: Harry N. Abrams, Inc.

MORRIS, D. (1985). *Bodywatching.* New York: Crown Publishers Inc.

PRATER, G. S. (1992). Child welfare and African-American families. In N.A. Cohen (Ed.), *Child welfare: A multicultural focus* (pp. 84–106). Boston: Allyn & Bacon.

RATHUS, S. (1993). *Psychology* (5th. ed.). Fort Worth, TX: Harcourt Brace Jovanovich.

ROGERS, C. (1961). *On becoming a person.* Boston, MA: Houghton Mifflin.

SATIR, V. (1988). *The new peoplemaking.* Mountain View, CA: Science and Behavior Books, Inc.

SHAKESPEARE, W. (1971). Hamlet, Prince of Denmark. In *The complete works of William Shakespeare.* London: The Hamlyn Publishing Group Limited.

WEBSTER'S NEW WORLD DICTIONARY. (1984). New York: Warner Books.

WORTHINGTON, R. L. (1992). *Education 165L Introduction to counseling psychology skills laboratory: A manual for teaching assistants.* Publication of the University of California, Santa Barbara, Office of Instructional Development.

Index

TO THE OWNER OF THIS BOOK:

I hope that you have enjoyed *Active Communication*, as much as I have enjoyed writing it. I'd like to know as much about your experiences with the book as you care to offer. Only through your comments and the comments of others can I learn how to make *Active Communication* a better book for future readers.

School: _____

Your instructor's name: _____

1. What I like most about this book is: _____

_____

_____

2. What I like least about this book is: _____

_____

_____

_____

3. The name of the course in which I used this book is: _____

_____

_____

4. Were all of the chapters of the book assigned for you to read? _____

If not, which ones weren't? _____

_____

5. In the space below, or on a separate sheet of paper, please write specific suggestions for improving this book and anything else you'd care to share about your experience in using the book.

_____

_____

_____

_____

_____

Optional:

Your name: _____ Date: _____

May Brooks/Cole quote you, either in promotion for *Active Communication* or in future publishing ventures?

Yes: _____ No: _____

Sincerely,

Matthew Westra

FOLD HERE

FOLD HERE

Brooks/Cole is dedicated to publishing quality publications for education in the human services fields. If you are interested in learning more about our publications, please fill in your name and address and request our latest catalogue, using ths prepaid mailer.

Name: _____

Street Address: _____

City, State, and Zip: _____

FOLD HERE

FOLD HERE